The Cultural Complexity of Carbon

This volume discusses the transformational role that carbon – both as a concept and as a distinct set of material forms and effects – has come to play in social and cultural life.

As a proxy for greenhouse gas emission data, carbon has grown to become a phenomenon that can no longer be accounted for solely within the technoscientific vocabulary of climate scientists. *The Cultural Complexity of Carbon* examines the extent to which our knowledge of carbon affects the way that human beings relate to each other and to the climate and/or the environment. It draws on case studies from a diverse range of topics, including peatland restoration, religion and energy systems, to explore questions that have so far been under-explored in the current literature. These questions include whether the recognition of carbon's role in climate change leads to an incremental adaptation of lifestyles or to cultural or existential transformations, but also more concretely how carbon is made meaningful, and how these meanings are attached to ideals of cultural change or continuity. Spanning multiple perspectives and disciplinary positions, this volume provides a go-to point for the next generation of ethnographic studies of carbon and climate change. It cuts across what has hitherto been largely separate literatures in anthropology, geography and sociology to provide a meta-level orientation to how contemporary narratives of the role of carbon are being told.

By addressing the intimate social and cultural changes that stem from humanity's involvement with its natural and climatic resources, this volume is of interest to students and scholars of climate change within the social sciences and environmental humanities.

Steffen Dalsgaard is Professor in Anthropology of Digital Technology at the IT University of Copenhagen, where he is heading the interdisciplinary Center for Climate IT.

Andy Lautrup is an ethnographic researcher studying youth climate activism in Scandinavia, drawing on insights from anthropology, science and technology studies and cultural theory.

Katinka A. Schyberg is an anthropologist by training and holds a PhD from the IT University of Copenhagen, where she is associated with the Technologies in Practice research group (TiP).

Ingmar Lippert, Anchor Lecturer of Goethe University Frankfurt's STS programme at the Institute of Cultural Anthropology and European Ethnology, sustains a research focus on environmental governance and its digital reconfigurations.

Routledge Advances in Climate Change Research

Climate Security
The Role of Knowledge and Scientific Information in the Making of a Nexus
Matti Goldberg

COVID-19 and Climate Change in BRICS Nations
Beyond the Paris Agreement and Agenda
Edited by Ndivhuho Tshikovhi, Andréa Santos, Xiaolong Zou, Fulufhelo Netswera, Irina Zotona Yarygina and Sriram Divi

Social Transformation for Climate Change
A New Framework for Democracy
Nicholas Low

Exploring Climate Change Related Systems and Scenarios
Preconditions for Effective Global Responses
Jeremy Webb

Resolving the Climate Crisis
US Social Scientists Speak Out
Edited by Kristin Haltinner and Dilshani Sarathchandra

Climate Politics in Populist Times
Climate Change Communication Strategies in Germany, Spain, and Austria
Mirjam Gruber

Confronting Climate Coloniality
Decolonizing Pathways for Climate Justice
Edited by Farhana Sultana

The Cultural Complexity of Carbon
Green Transformations in Contemporary Society
Edited by Steffen Dalsgaard, Andy Lautrup, Katinka A. Schyberg and Ingmar Lippert

For more information about this series, please visit: www.routledge.com/Routledge-Advances-in-Climate-Change-Research/book-series/RACCR

The Cultural Complexity of Carbon

Green Transformations in Contemporary Society

Edited by Steffen Dalsgaard, Andy Lautrup, Katinka A. Schyberg and Ingmar Lippert

LONDON AND NEW YORK

First published 2025
by Routledge
4 Park Square, Milton Park, Abingdon, Oxon OX14 4RN

and by Routledge
605 Third Avenue, New York, NY 10158

Routledge is an imprint of the Taylor & Francis Group, an informa business

© 2025 selection and editorial matter, Steffen Dalsgaard, Andy Lautrup, Katinka A. Schyberg and Ingmar Lippert; individual chapters, the contributors

The right of Steffen Dalsgaard, Andy Lautrup, Katinka A. Schyberg and Ingmar Lippert to be identified as the authors of the editorial material, and of the authors for their individual chapters, has been asserted in accordance with sections 77 and 78 of the Copyright, Designs and Patents Act 1988.

With the exception of the Introduction, no part of this book may be reprinted or reproduced or utilised in any form or by any electronic, mechanical, or other means, now known or hereafter invented, including photocopying and recording, or in any information storage or retrieval system, without permission in writing from the publishers.

The Introduction of this book is freely available as a downloadable Open Access PDF at http://www.taylorfrancis.com under a Creative Commons Attribution-Non Commercial-No Derivatives (CC BY-NC-ND) 4.0 license.

Any third party material in this book is not included in the OA Creative Commons license, unless indicated otherwise in a credit line to the material. Please direct any permissions enquiries to the original rightsholder

Trademark notice: Product or corporate names may be trademarks or registered trademarks, and are used only for identification and explanation without intent to infringe.

British Library Cataloguing-in-Publication Data
A catalogue record for this book is available from the British Library

Library of Congress Cataloging-in-Publication Data
Names: Dalsgaard, Steffen, editor. | Lautrup, Andy, editor. | Schyberg, Katinka A., editor. | Lippert, Ingmar, editor.
Title: The cultural complexity of carbon : green transformations in contemporary society / edited by Steffen Dalsgaard, Andy Lautrup, Katinka A. Schyberg and Ingmar Lippert.
Description: Abingdon, Oxon ; New York, NY : Routledge, 2025. | Series: Routledge advances in climate change research | Includes bibliographical references and index.
Identifiers: LCCN 2024050363 (print) | LCCN 2024050364 (ebook) | ISBN 9781032764856 (hbk) | ISBN 9781032764863 (pbk) | ISBN 9781003478669 (ebk)
Subjects: LCSH: Atmospheric carbon dioxide--Social aspects--Case studies. | Carbon--Social aspects--Case studies. | Climate change mitigation--Social aspects--Case studies. | Climatic changes--Social aspects--Case studies. | Human beings--Effect of climate on--Case studies.
Classification: LCC QC879.8 .C85 2025 (print) | LCC QC879.8 (ebook) | DDC 304.2/8--dc23/eng/20250321
LC record available at https://lccn.loc.gov/2024050363
LC ebook record available at https://lccn.loc.gov/2024050364

ISBN: 978-1-032-76485-6 (hbk)
ISBN: 978-1-032-76486-3 (pbk)
ISBN: 978-1-003-47866-9 (ebk)

DOI: 10.4324/9781003478669

Typeset in Times New Roman
by SPi Technologies India Pvt Ltd (Straive)

Contents

List of illustrations vii
List of contributors viii
Preface and acknowledgements xi

Introduction: Carbon and culture change 1
STEFFEN DALSGAARD

1 Renewable energy communities for energy-poor households: Policy mobility challenges in urban and peri-urban European contexts 25
SIDDHARTH SAREEN AND BÉRÉNICE GIRARD

2 Carbon footprint calculators and behaviour change 46
QUENTIN GAUSSET

3 Configuring the carbon farmer: Emerging practices of carbon accounting and biochar engagements in Danish agriculture 65
INGE-MERETE HOUGAARD

4 The footprint of anarchy: Counting carbon in the Church 86
KATINKA A. SCHYBERG

5 Carbon in Chinese notions of ecological civilization: Policy of quantification or philosophy of promise? 110
CHARLOTTE BRUCKERMANN

6 The social life of peat carbon and peat frontier making: An ethnographic study of peatland restoration in Central Kalimantan 132
ANU LOUNELA

7 La#oma's pre- and post-carbon landscape: The ont*-politics of a vanished village 154
INGMAR LIPPERT

8 Scaling the world through carbon: Discursive decoupling, unity and climate responsibility in Stavanger, Norway 185
ANDY LAUTRUP

Index 204

Illustrations

Figures

5.1	Poster for Fuzhou Eco-civilization campaign. Poster caption: 绿色城市和谐家园: 共创生态手牵手，生态文明心连心	111
	"Green City, Harmonious Homeland: Hand in hand to create our common ecology, heart to heart in building ecological civilization." (author's translation)	
6.1	Villager passing the dam on a boat during the restoration assessment in February 2019.	146
6.2	Travelling along the newly cleaned channel in the village after the wildfires in the fall of 2019.	148
7.1	Archive: view of its single exhibiting room.	163
7.2	Map: Historical view of Lakoma.	163
7.3	Map: Contemporary view of Lakoma in its regional context.	164
7.4	Archive: Infosauger and the mapped floor around Lakoma.	165
7.5	Left before the bridge across the New Hammergraben to La#oma: 13 white crosses and the artistic windmill.	168

Tables

1.1	Comparison between Solar for All and Eurosolar4All.	34
1.2	Specificities of each context.	35
1.3	Departures from basic design.	35

Contributors

Charlotte Bruckermann's ongoing research focusses on the intersection of social reproduction, economic accumulation and ecological transformation in China, with a particular emphasis on carbon. Her ethnographic fieldwork spans several key carbon hotspots in China, where she explores how carbon's cultural, economic and ethical value shapes the construction and imagination of Chinese ecological civilization. In her research, she explores the multifaceted role of carbon in local and national discourses on sustainability, development and ecological stewardship. She is the author of *Claiming Homes: Confronting Domicide in Rural China* (Berghahn Books, 2020) and co-author with Stephan Feuchtwang of *The Anthropology of China: China as Ethnographic and Theoretical Critique* (Imperial College Press, 2016). Her recent publications, including various articles on kinship, care, environment and ethics in China, critically engage with the evolving juncture between carbon economies and ecological governance.

Steffen Dalsgaard is a Professor in Anthropology of Digital Technology at the IT University of Copenhagen, where he is heading the interdisciplinary Center for Climate IT. He has previously conducted long-term research in Papua New Guinea and is now exploring how "carbon" has become a central value form in the cultural transformation of Western societies. He is currently directing a number of research projects, including the ERC-funded DecouplingIT.

Quentin Gausset received his PhD in Anthropology from the Free University of Brussels and is Associate Professor at the Department of Anthropology of the University of Copenhagen. His research focusses on the sociocultural aspects of environmental management and on the collective dimension of environmental behaviour. He is currently leading a research project studying green communities in Denmark.

Bérénice Girard is a researcher with the French National Research Institute for Sustainable Development. Her research sits at the crossroads of environmental sociology, sociology of the State and South Asian Studies. Her work has focussed on the management of the Ganges River, on energy changes in urban and urbanizing localities of Northern India and on the transition to solar energy in India. Her PhD in sociology at EHESS Paris won the GIS Asie 2020 PhD Award. She has held postdoctoral fellowships at Gustave Eiffel University (2018–2020), the Centre de Sciences Humaines (2021) and the University of Stavanger in Norway (2021–2023).

Inge-Merete Hougaard holds a PhD in Political Ecology from the Department of Food and Resource Economics, University of Copenhagen, and has a background in International Development Studies and Public Administration. Her main research interests revolve around resource rights, landscape change, state-making and climate politics, with fieldwork experience in Denmark and Colombia. Her current research at the Department of Anthropology, University of Copenhagen, focusses on the role of storytelling in multispecies landscape co-creation in the context of green transition in Danish agriculture. This chapter draws on her previous research on the role of carbon dioxide removal in Danish climate politics.

Andy Lautrup is an ethnographic researcher studying youth climate activism in Scandinavia, drawing on insights from anthropology, science and technology studies and cultural theory. They received their PhD from the IT University of Copenhagen in December 2022 and are currently holding a postdoc position at the Department of Arts and Cultural Studies at the University of Copenhagen. Here they research the climate crisis as a crisis of care, focussing on youth climate activism in Denmark, and teach interdisciplinary courses in the field of environmental humanities.

Ingmar Lippert, Anchor Lecturer of Goethe University Frankfurt's STS programme at the Institute of Cultural Anthropology and European Ethnology, sustains a research focus on environmental governance and its digital reconfigurations. He has been studying Lusatian naturescultures from 2018 till 2024, subsequent to a decade of research on corporate environmental accounting. Ingmar has published widely on corporate climate-relations, carbon management accounting, sustainability governance, digitalization and numbers as well as on STS methods and methodologies. Collaborative work includes the special issues "Environmental Management as Situated Practice" (Geoforum, 2015), "After Numbers? Innovations in Science and Technology Studies" Analytics of Numbers and Numbering" (Science & Technology Studies, 2018) and "Methodography of Ethnographic Collaboration" (Science & Technology Studies, 2021).

Anu Lounela is an anthropologist and university researcher in Global Development Studies and Social and Cultural Anthropology at the Faculty of Social Sciences, University of Helsinki. Combining environmental anthropology and political ecology, as well as multispecies and landscape approaches, she has published widely on frontier making, human–nonhuman relationships, and riverine and wetland life, as well as wildfire occurrences and their impact in the Indonesian side of Borneo. Her research on water and vulnerability

(Academy of Finland) explored historically informed socionatural relations and social forms and transformations of wetlands. She is currently involved in the project "Repair and Responsibility in ruined environments" (REPAIR), funded by Kone Foundation (2024–2027). She was a university researcher for the research project "Water and Vulnerability in Fragile Societies" (Academy of Finland, 2018–2022) and Principal Investigator for the research project "New regimes of commodification and state formation on the resource frontier of Southeast Asia" (Kone Foundation, 2018–2023). Her research interests include environmental anthropology and political ecology, socionatural landscapes, wetlands, water and fire, multispecies relations, social values and state formation in Kalimantan, Java and Indonesia.

Siddharth Sareen is a research professor at the Fridtjof Nansen Institute, Lysaker, Norway. He researches the governance of energy transitions, has worked in seven countries, and is an interdisciplinary environmental social scientist. His research focusses on social and environmental equity and justice. He was awarded the Nils Klim Prize 2024 and IK Lykke Prize 2023 for research excellence. He has ethnographic experience along the energy value chain in India, Portugal and Norway and serves on the Empowered Futures Research School and Letten Prize boards.

Katinka A. Schyberg is an anthropologist by training and holds a PhD from the IT University of Copenhagen, where she is associated with the Technologies in Practice research group (TiP). Her doctoral research focussed on environmental engagements in the Danish People's Church, and in her PhD thesis she explored how this well-established national church navigates tensions between continuity and change in the process of performing a green transition. Schyberg has previously conducted research on Evangelical Christianity in the United States and takes a general interest in cosmology, religion and conceptions of nature and technology.

Preface and acknowledgements

The foundation for this volume was established about a decade ago with the idea that carbon had to be approached as "multiple" due to the many cultural perspectives on what carbon "means" across different social and cultural groups. The idea led to research proposals, which eventually led to a research grant, which then again led to a research project, which pursued questions about how people in different sociocultural contexts interpret data about climate change and translate them into action and/or ethical reflections. Along the way, it became important to retain a focus on how carbon entails a specific cultural imagination of the world as somehow united in being impacted by one specific element, albeit in vastly different ways. It also became important to accept the many different connotations and interpretations that this gives rise to.

In spring 2023, the book's editors organized a symposium at the IT University of Copenhagen to wrap up the project's findings and to get a wider sense of how our ideas resonated with those of the different types of research, which over the years have treated carbon as a social and/or cultural artefact. The symposium brought together scholars from various scholarly fields and traditions who all share an interest in the sociocultural dimensions of climate change and the role of carbon. All contributors participated in this symposium, and so did Katja Müller, Anders Blok, Simon Ulrich and Kari Marie Norgaard, who we were grateful to have as participants and commentators on the papers and in this way helped shape the perspectives and discussions that emerged from the present work.

The funding for much of the research included in this volume came from Steffen Dalsgaard's Sapere Aude Grant from the Independent Research Fund Denmark for the project SOCiocultural CARbon (SOCCAR). Many more people than can be mentioned here have given input along the way. Most importantly, we want to mention the project's advisory board, which consisted of Paige West, J.C. Salyer, Patrick Bigger, Thomas Hylland Eriksen, Daniel Seabra Lopes, Hannah Knox and Joel Robbins, and the ITU's Research Support unit, especially Pia Kystol Sørensen. We are also grateful for the careful editing and proofreading of this manuscript by Jeremy Gunson and to Thomas Murphy for diligent formatting.

Introduction

Carbon and culture change

Steffen Dalsgaard

Introduction

Since the 1990s, carbon has emerged as a new social and cultural resource. In a variety of forms, ranging from government regulation of emissions, via the branding of services and goods as "carbon neutral," to the awareness of the climatic costs of individual consumption, carbon has gained prominence in the everyday lives of a diversity of actors across differences of class, profession, gender, age, nationality, technical capacity and much more. As a shorthand for greenhouse gas emissions, carbon has grown to become a data phenomenon that can no longer be accounted for solely within the technoscientific vocabulary of climate scientists, nor can it any longer be treated as merely an economic externality to human modes of production. A variety of new techniques and infrastructures have provided calculative and representational possibilities for governments, corporations, organizations and individuals interested professionally or personally in knowing their "carbon footprint" or in (global) comparisons of emissions. The possibilities engendered by calculative and informational spaces have set up carbon as a "metric of the human" (e.g. Whitington 2016) and as a datafied and informational object that can be transacted as credits or allowances for productive or consumptive action (e.g. Knox-Hayes 2013). Thus it has, in the words of geographer Gavin Bridge, fast become "a common denominator for thinking about the organization of social life in relation to the environment" (2010:821). As a new value form, carbon has entered individual and collective imaginaries across the globe. Yet, in this volume, we contend that the ways it has done so are by no means uniform.

While it is certainly pertinent to ask to what extent the metric of carbon contributes to the desired and intended goal of mitigating climate change through a global reduction of emissions, *a more pressing anthropological and social scientific question is to what extent this new status of carbon as a datafied object generates any change in human social and cultural practices.* In other words, what kinds of specific – intentional or unintentional – changes or "transformations" does carbon make in people's lives across different societal and cultural domains? This volume discusses the transformational role that "carbon" – both as a concept and as a distinct set of material forms and effects

– has come to play in social and cultural life. It is about how carbon is made meaningful, and how its meanings are attached to ideals and values of social and cultural change but conversely also to cultural continuity.

This role of carbon as involved in cultural change is one that extends well beyond what in public and political discourse have come to be known in vague terms as "green transitions" or "green transformations" (e.g. Scoones et al. 2015), whose abstract end is frequently – but not always – accounted for in terms of reducing greenhouse gas (carbon) emissions. It is beyond the scope of this volume to go into discussions of the nuances of the political narratives of green transformations or transitions.[1] We are instead interested in what appears if we look beyond the corporate claims of being "carbon neutral" or reaching "net zero" emissions and beyond governmental claims of being on target to make promised reductions. Carbon emissions are rising across the globe, despite promises of green transformations and accounts of the formation of "green subjects" and "climate conscious consumers." For this reason, there is ample reason to look beyond the "value" of carbon in markets and in the economic calculations. This entails looking specifically at the role of carbon itself as an object, which can be identified almost anywhere and everywhere because of its physically – and increasingly culturally – pervasive nature (see Dalsgaard 2013; Ervine 2018). That is, while carbon is physically present in multiple forms and places, it is also a cultural resource that covers multiple meanings, both in terms of how humans are beginning to interpret their role vis-à-vis nature and the planet differently and in terms of what carbon as a concept or a value form affords in terms of actions and imaginaries under different social, cultural and infrastructural circumstances.

The questions addressed in this volume

The impact that carbon as a metric and as a value form has had – or has failed to have – in driving green transformations raises several interrelated questions. In this volume, we attend in particular to the following that relate to how carbon is socially or culturally important: To what extent does the knowledge of carbon affect the way that human beings relate to each other and to the climate and/or the environment – both individually and collectively? To what extent does the knowledge of carbon (as data or as information) allow for cultural identities to emerge as more or less "carbon counting" or "green" subjects? Does the recognition of carbon's role in climate change lead to an incremental adaptation of lifestyles or rather to cultural or existential transformations? Or is it met with socially organized denial (Norgaard 2011)? For carbon to have any actual agency in generating change, can we expect it to be visible in everyday life practices and decisions that affect how the lives of human beings are otherwise organized – politically, legally, religiously and economically, for example? Given the vastly different effects and experiences of climate change globally and the vastly different capacities and constraints in how individuals or collectivities respond to climate change (e.g. Roncoli et al. 2009), it is also pertinent to ask how carbon is differentially valued in cultural terms. Carbon is truly a global phenomenon, but we do not assume that its meaning and value are understood and interpreted in similar ways across the world. But how, then, do different people relate to carbon in different ways?

These questions demand attention to how human beings encounter carbon in their daily lives, whether as a material object, as embedded in a variety of infrastructures, or as discourse – as something we "talk" about but cannot really touch or feel, except through the diverse practices that we associate with emissions. Since our encounters with carbon imply ways of valuing it, we also need to pay attention to culture, since what and how things are valued is inherently cultural. By focussing on culture and its role in social relations, we aim to contribute to social science discussions of climate change with a deeper understanding of carbon's role in contemporary society, but we also take the concept of culture more seriously than other volumes, which claim to address carbon or its related institutions as somehow "cultural," but either delimit their focus to institutions of green transitions in the Global North (Bulkeley, Paterson and Stripple 2016) or do not define their concept of culture (e.g. Knox-Hayes 2016).

In many anthropological definitions, culture is understood as a matter of how relationships, meanings and interpretations of reality are seen as similar or different from one another. Culture is how human actors categorize the world, which is not merely about the everyday lives of citizens, consumers, individuals or communities as bounded entities, but about the way these actors relate to one another and to each other's values and knowledge forms, including their mutual entanglements, complexities and contradictions. This includes how the meanings these actors generate relate in different ways to the cultural influences that stem from the scientific, political, economic or legal domains that in themselves are by no means homogeneous but nonetheless have a huge impact on climate change mitigation and the different valuations *of* but also *with* carbon. In this respect, the introduction presents the contributions to the volume and frames these through a focus on the diverse meanings ascribed to carbon in different cultural contexts. We also discuss how cultural meanings and social relations change or stay the same in the process of making carbon valuable across a variety of contexts. We claim that (a) carbon as an object can be treated as a fundamental part of contemporary culture and that (b) it lends itself to being studied as culture. That is, it can be studied by qualitative and ethnographic means.

To address carbon's transformational potential as a cultural artefact – that is, how it is implicated in culture change – it is necessary to discuss how carbon has emerged as valuable in a variety of different ways, but also how it has made old valuables valuable in new ways. For example, is the value of one's forest the same if under new carbon metric standards of value, trees can be sold not as wood but as carbon offsets? And how do people in Norway value oil when its importance has moved from that of being the enabler of welfare to a substance that, through its combustion, damages the welfare of the entire planet (Lautrup, this volume)? This difference in part stems from how carbon means different things to different people in different contexts. For example, outside of the disciplinary perspectives of atmospheric chemistry and climate sciences, carbon exists in governance and policies (from international to municipal levels) and in financial circles and infrastructures (as credits, permits, etc.). Its existence in the latter allows market actors to invest and hope to make a profit while simultaneously believing that their work will positively impact the way that carbon – in the sense of greenhouse gas emissions – is reduced elsewhere. Yet, while these domains have played a critical role in transmitting ideas about how and why carbon could or should be valued to a diversity of other domains and publics, carbon's meanings cannot be confined to those ascribed to it in science, in policies or in market institutions. Our focus on culture addresses this "elsewhere," where carbon is much more than a greenhouse gas, a target in policy directives or a financial asset. Carbon exists in a plethora of other social and cultural contexts. For example, when carbon is identified in the form of "footprints," it becomes morally indicative of how people want or ought to live their lives (e.g. Paterson and Stripple 2012; Dalsgaard 2022). Carbon has furthermore become institutionalized in corporate or governmental accounts and inventories, but also in everyday practices associated with "low-carbon" living (e.g. Whitmarsh and O'Neill 2010; Lovell 2015). Even where carbon is "stored" in natural environments (e.g. forests, fields, etc.), these environments cannot solely be reduced to "carbon sinks." They are often intimately entangled with human agency relating to resources that are owned and managed by someone and for whom they carry specific meanings (e.g. Mahanty et al. 2012; Paladino and Fiske 2017; Bruckermann, this volume; Lounela, this volume). In this way, carbon – no matter where it is found or in what form it is identified – is entangled in social and cultural webs of significance.

In the remainder of this introduction, we first attend to carbon as a range of diverse phenomena in human lives and, second, to the way that it can be approached as culture. Then we present a discussion of its potential for generating or promoting change, before turning to the different contributions to this volume and how they provide unique perspectives on the topic of carbon. In sum, in the introduction, together with the volume as a whole, we demonstrate how paying attention to carbon as a cultural artefact allows for a more profound appreciation of when, how and why carbon enables (or sometimes disables) change.

How the social sciences have approached carbon: a multitude of phenomena

There are many starting points for an exploration of the meanings and values ascribed to carbon. The following pages outline four topics that in our reading have dominated the social science literature on carbon. These are (a) carbon's role in "the green transition," (b) carbon as a metric, (c) carbon as a material object and (d) carbon in everyday life. There are multiple overlaps between them, and they must not be seen as mutually exclusive.

One starting point is, as mentioned above, how carbon is today regarded as central to what has variously been dubbed "the green transition" or "green transformation" of human societies. Calls for transitions to a "low-carbon economy" are evidently one of the key components of these imaginaries. The type of questions engendered when focussing on carbon's role in these processes lend themselves to exploration from the perspectives of political economy in a broad sense (Scoones et al. 2015). That is, how especially political scientists, geographers and economists have studied the ways that political and economic structures govern or are transformed in or by the economy of carbon. A central component of these perspectives is the study of the institutional processes behind the phenomenon of climate governance mechanisms and, in particular, carbon markets – both compliance and voluntary markets[2] (e.g. Helm and Hepburn 2009). Initial interest from economists followed the lines of market design in response to the implementation of the Kyoto Protocol's flexible mechanisms (e.g. Grubb 2003; Yamin 2005) and the making of the regulatory or compliance markets typically governed by states or international organizations. These concrete markets, such as the European Union's Emissions Trading Scheme, have received much attention (e.g. Bigger 2016; Wettestad and Guldbrandsen 2018), but the processes of establishing markets that failed to come into place have also been studied (Pearse 2018). From there, the more obvious political angles have become integrated into discussions of the formation of a new "carbon economy" (Bridge 2010; Newell et al. 2012) or "climate capitalism" (Newell and Paterson 2010). These include discussions of how a variety of actors – and especially markets – are deployed to govern climate change mitigation (e.g. Bulkeley and Newell 2010; Knox-Hayes 2016; Ahonen et al. 2022), how public sector institutions have adopted their own emission reduction goals or projects (e.g. Rice 2010; see also Knox 2020) and how the market designs, including cap setting, offsets and allocation of allowances, have spillover effects and social consequences, even in those cases where they manage to deliver the emission reductions they promise (e.g. Goodman and Boyd 2011; Ervine 2018).

Another potential starting point is attending to the capacity of carbon to be deployed as a metric. Some of the scholarship in this field focusses on how individual techniques of the self are promoted, while others have studied corporate climate accountability, which is facilitated by management and accounting concerns and practices. Much of the literature addresses the technical measures and procedures that define how carbon can be identified across different fields and how it is commoditized, priced and valued technically in different contexts (e.g. Bowen and Wittneben 2011; Yocum 2016; Ehrenstein and Muniesa 2013; Gifford 2020). Other contributions dive into the supporting practices and infrastructures of carbon metrics, from excel sheets and meeting notes to apps, digital repositories and devices that set up frameworks for participation and action (e.g. Marres 2011; Lippert 2015, 2018). The design of the metrics, as well as their material concretization in infrastructures, has been demonstrated to be both influenced by corporate or technical "cultures" of accounting and accountability (e.g. Lovell and MacKenzie 2012) and by definitions of carbon that sometimes appear almost arbitrary but nevertheless come across as politically useful for accounting "proxies" for emissions (see Lippert 2018; Dalsgaard 2024). Rather than accepting their status as purely technical, critical social science scholarship has pointed out how the modes of standardization and definitions of what counts as emissions contain normative, political and disciplinary power (e.g. Gupta et al. 2012), and indeed carbon metrics are frequently thought of as facilitating government intervention and commoditization (see Knox 2020). Carbon metrics also impact the way that organizations and individuals relate to or even enact "the climate" as a datascape through accounting (Lippert 2015). More recent contributions to these discussions highlight the role played by corporate attempts to construct footprints, for example, through the scopes invented by the GHG Protocol (e.g. Walenta 2021). Analysis of carbon's function as a metric is especially inspired by science and technology studies and governmentality theory. Within these approaches, carbon is consequently seen to impact human social and cultural choices and to engender new "green" subject formations (e.g. Lovell and Liverman 2010; Paterson and Stripple 2010; Lövbrand and Stripple 2011).

A key component in the construction of carbon as a driver of subjectification is its commoditization as credits or allowances. This relies not only upon techniques of counting and accounting but also frequently upon the objectification of carbon into material and often physical forms that can be counted and accounted for. Due to the mechanisms of the Kyoto Protocol and the technical rendering of different greenhouse gases as commensurable (e.g. Mackenzie 2009; Dalsgaard 2013), carbon numbers (as credits or allowances) are made in a multitude of physical locations, infrastructures, technologies and techniques in the Global North as well as the South, where carbon is identified in, for example, land use changes such as agricultural production, biofuels, forest tenure and the management of carbon sinks (e.g. Leach and Scoones 2015; Paladino and Fiske 2017), and in practices related to energy saving such as improved cookstoves (e.g. Wang and Corson 2015) or electricity systems (e.g. Boyer 2019; Howe 2019; Sareen and Müller 2023). A recent concretization is the development of new "carbon capture" technologies and the subsequent infrastructures that allow concrete carbon dioxide to be circulated, stored or utilized (e.g. Günel 2016; Buck 2022). The focus on forests has been especially scrutinized by academics due to the attention that the ongoing negotiations over the REDD+ programme[3] – as well as its effects on forest and land tenure in the Global South – has received in international climate change debates (e.g. Corbera and Schroeder 2011; Murdiyaso et al. 2012; Pascoe 2018; Greenleaf 2020). In parallel, carbon numbers are also identified by those searching for credits to buy. They find them in devices for "everyday carbon accounting" (e.g. Marres 2011; Lövbrand and Stripple 2011; Gausset, this volume).

A final point of departure is then how carbon is identified in everyday life practices or discourse – not as a commoditized "object," but as something related to actions performed on a daily basis and to the semiotic and metaphorical references that inform and ascribe meaning to those actions. That is, how carbon is associated with specific norms and ways of living, acting or consuming (e.g. Lovell et al. 2009; Lindman et al. 2013), and how carbon is attached to the cultural meanings of things and actions at work or at home, for example. This includes the metaphorical use of carbon footprints or other linguistic "carbon compounds" to reflect upon climate impacts (e.g. Nerlich and Koteyko 2009; Girvan 2018), but also the anthropological attention given to the choice between different types of everyday practices of governance or consumption in response to deteriorating environmental conditions (e.g. Knox 2020) and how these practices differ across a variety of distinctions such as residential location, gender and age (e.g. Underwood and Fremstad 2018; Albert et al. 2020). Associations with specific ways of consuming are semiotic, but they are often also normative (e.g. Dalsgaard 2022), and one line of analysis pursued by only a few scholars has been the Bourdieu-inspired discussions of how carbon altogether relates to "green" forms of capital, status or social distinctions (e.g. Horton 2003; Lippert 2013).

Carbon as cultural

What the above approaches have in common is how they all attempt to answer how carbon has become "valuable." Whereas previously, it was evident how the value of carbon stemmed from its physical and chemical role as a component in fossil fuels that allowed societies to grow based on the consumption of energy (e.g. Mitchell 2011; Appel, Mason and Watts 2015; Di Muzio 2015), one can today make the more general observation that carbon has become valuable in its own right due to its climatically negative value as a key component in the production of greenhouse gases. The big question is how does this climatic value manifests itself in a diversity of social contexts? "Climate change" is itself valued differently by different individuals and collectivities. Even if victimhood and vulnerability can be debated (Hughes 2013), some social groups surely have the resources to adapt to – or even gain from – the subsequent sociomaterial changes, while others – the majority – seem positioned to lose (see Klein 2014; Singer et al. 2016). Climate change – and with it, carbon – thus translates easily into value in economic and material terms. Yet economic and material terms are also cultural terms. Livelihoods and the many different actions and objects associated with carbon emissions are all valued in relation to things other than "just" climate change and the costs implied in efforts to avert it. Economic value is but one of several cultural "values" against which something can be assessed as valuable, and the materialities that come to matter can themselves be defined as culturally constituted. An example frequently referred to by the anthropologist Marshall Sahlins was that no ape would be able to tell the difference between holy water and distilled water because they are chemically identical (1999:400). Similar struggles over the valuation of carbon take place every day when corporations and consumers alike are asked to choose between emissions that, from some perspectives – such as that of business accounting – appear identical, but in terms of other social or cultural meanings are entirely different – not the least in the effects that are derived from this difference, which can be either of semiotic or of moral value, or most likely both. A perspective on culture inspired by Sahlins' work (e.g. 1976; 1999) would suggest that such differences in meaning can only be established with reference to culture as a primary and autonomous entity in relation to the materialist, utilitarian or "economistic" rationalities that have otherwise today come to dominate, especially through what Scoones, Newell and Leach (2015) refer to as the technocentric, marketized or state-led narratives of transformation. That is, there is much more to human existence and to what we value than that which can be explained by political or economic rationalities. Of course, there can be no anthropological text about culture without caveats. The culture concept (or concepts, since there is hardly any agreement on how to define culture) faced multiple critiques towards the end of the previous century (see Brightman 1995). However, it still exists, and it remains in operation in anthropology and beyond for those trying to find a way to address deeper patterns and inconsistencies in human meanings, processes, behaviours and practices.

Drawing upon a social anthropological perspective, we also emphasize how culture must be understood as emerging from practice as a relational concept rather than referring to essentialist characteristics (e.g. Strathern 1995). It is about the practice of making social and cultural distinctions between self and other, as much as it is about defining subject and object, nature and culture "good and bad" emissions or "good and bad" climate actors. No matter the right balance between (or rather integration of) these distinctions, a more nuanced understanding of *value* than that defined in economic terms is crucial if we are to understand the breadth of responses to climate change and the management of emissions. To approach the practical relationality of culture, here we take our cue from the anthropological critique of economic thinking presented by Sahlins (1976), but also from more recent attempts to revive an anthropological concept of value (Graeber 2001; Otto and Willerslev 2013). In particular, David Graeber's work on value provides a useful lens for how to address the contrast between different understandings of value, and how value is an innately cultural phenomenon, which in anthropological discussions interchangeably refers to economic value (related to markets, price, etc.), moral value (e.g. social norms) or semiotic value (related to linguistic usage within a structure of meaning). First, what is of value and how value is defined are closely interrelated to discussions of culture (see, e.g. Otto and Willerslev 2013). Second, the topic of carbon seems easily adaptable to Graeber's discussion of different anthropological approaches to value because of carbon's ephemeral nature in everyday life and because of the multiple ways that carbon is valued and regarded as valuable – from the spheres of science and markets to norms and everyday practices. Relying on these sources of inspiration helps us frame the chapters of the present volume in relation to how carbon is identified and valued across such a wide variety of human relationships. Instead of seeing culture as a separate realm "exogenous" to either nature (climate) or the economy, inspiration from the thinking of Sahlins and Graeber points to how both these entities are fundamentally cultural.[4]

Graeber (2001) grounds his discussion and theorization of value in the creativity of human action and innovation (Otto 2024: 30; Graeber 2001: 259). As a self-proclaimed anarchist, Graeber's hope was to free human action to fulfil "its creative potential of imagining different values and a different society" (Otto 2024: 30). The 1997 Kyoto Protocol did indeed place carbon at the centre of such a transformative aspiration for imagining and creating a new type of value (if not necessarily a different global society), but this was an aspiration originally envisaged by policymakers and economists, who built upon ideals of neoliberal globalization and financialization when they allowed carbon in whatever forms it could take to be commodified and marketized (see Lohmann 2010). This is almost certainly not what Graeber had in mind, not the least in the wake of the great financial crisis and the prioritization of bailouts for finance (2013:235). Graeber's contribution to discussions of value is rather his emphasis on the need to assess the value of actions (or objects, one might add), not only in relation to what they are imagined to be at a specific abstract moment but also in relation to what they have the capacity to become or to generate in practice in different cultural settings (2001:254). From this perspective, carbon emissions become about *potential* – of what may or may not be realized. The role of counterfactuals in carbon accounting is a case in point (e.g. Lohmann 2011; Ehrenstein and Muniesa 2013). Calculating or just identifying carbon as a value form proposes a distinct view of a *future* to be attained, but often one that is unclear to a lot of the people involved in terms of what this might entail, which some of the contributions here also touch upon (see Bruckermann, Hougaard, Lautrup, Lounela and Schyberg, all in this volume). At other times, this valuation of carbon is about establishing a narrative that apportions present-day responsibility for the past (Lippert, this volume). Emphasizing how carbon is cultural in this way ties into the idea that culture is practice and how all the different sectors, organizations and institutions that deal with carbon are culturally constituted – and so in turn they constitute the value of carbon differently through their particular organizational forms, systems, performances and processes (cf. Best and Paterson 2010: 13f).

Culture change

If carbon plays an important social and cultural role in contemporary societies, it is pertinent to ask what forms of transformational potential it might hold. What forms of change is carbon seen to enable or generate through the different forms it takes? And what forms does it disable? In other words, how can we approach the changes or continuities engendered by carbon as *cultural* change or continuity, with carbon as a new value form at the centre?

Putting carbon at the centre of discussions of change gives us an opportunity to see what happens when a new form of value is introduced or emerges in social and cultural life; it allows us to study the effects of enumerations and valuations that cut across a diversity of personal and organizational relationships in practice (cf. Dalsgaard 2013). The introduction of carbon as a financial asset and as a "unit," which individuals and collectivities can relate to and compare across their lives, is *meant* to generate change – albeit a change limited to distinct forms of production, consumption, governance and behaviour, which simultaneously is meant to ensure continuity by stabilizing the economic system and the climate in tandem. Few people would doubt that political and economic structures – including the making of new financial value forms – provide incentives, but how the changes in, for example, technologies, governance systems and markets impact culture and social practice in a way that not only changes extrinsic motivations but also changes norms, status forms and personal identities, or even how small-scale citizen-driven mobilizations can spread more broadly, remains an open question.

Policy initiatives oftentimes rely on technological changes that assume that we can achieve decarbonization without the need for people to change socially or culturally. Yet, there are also initiatives promoting small motivational interventions in individual attitudes and choices (Bulkeley, Paterson and Stripple 2016: 2). How sincere these are in generating norms for how a moral person – an individual – should reduce their climate impact has been profoundly debated. A significant body of Foucauldian-inspired theorization has anticipated the formation of green subjects based upon carbon accounting and other governmentality mechanisms (e.g. Paterson and Stripple 2010; Lövbrand and Stripple 2011). However, it is uncertain whether – or when – such mechanisms work. For example, Hannah Knox (2020: 93ff) has discussed how the city of Manchester discovered that, while theoretically possible, the implementation of consumption-focussed carbon footprinting turned out to be difficult to implement in practice. It was challenged both by the difficulty of tracing the minute details of how objects were made and by the ontological remapping and reconsideration of the roles of vast amounts of objects in social life. As with many of these consumption-based carbon footprint schemes, it is still uncertain the extent to which they can engender deeper cultural changes (either to the self or to collective norms and practices) or if they remain a form of lip service – a shallow "performance" of green identities (Horton 2003) or a discursive "simulation" (cf. Blühdorn 2007). Previous research by psychologists (among others) has pointed out that there is often a stark difference between the "intent" to make green changes and the actual climate impact of the actors in question (Moser and Kleinhückelkotten 2018). However, Quentin Gausset's research (this volume) demonstrates that the self may be configured communally through networks of neighbours and peers who motivate each other to first calculate emissions from their consumption and then subsequently support each other in reducing them. Qualitative and ethnographic work has a role to play here by analysing these changes not as a matter of individual preferences for either adoption or resistance to a new value form, but as part of social relationships between Self and Other and as part of semiotic distinctions that are shared even if they are not necessarily agreed upon.

Two important aspects of change are how quickly it occurs and its degree or "depth." Sometimes change may appear slowly and be gradual, while at other times it is sudden and radical. Given the urgency of the climate crisis, it is pertinent to consider the levels of change brought about by carbon. Some have stressed the need for individual change in behaviour and consumption practices as a necessary ingredient in mitigation, while other voices have focussed on what some would call deeper changes in perceptions, motivations and sociocultural values, or even (infra-)structural, societal or global changes to capitalism. It is worth remembering that carbon, as a value form, is meant to engineer forms of change that carry specifications. It must be "sustainable," "neutral," "net zero" or "climate friendly." It must help human collectivities and individuals reduce the greenhouse gas emissions they are responsible for. Somewhat paradoxically, if change has to be sustainable or involve "neutrality," per definition it also involves a certain amount of continuity – something that also might not change by being retained, reproduced or renewed. As argued by Sahlins in *Islands of History* (1985), the dichotomies of history and structure and change and continuity do not exist in opposition to each other. Culture functions as "a *synthesis* of stability and change, past and present, diachrony and synchrony." (Best and Paterson 2010: 144, emphasis in original). In other words, Sahlins argues that all change departs from existing forms. It is for this reason that a conceptualization of culture is valuable to the study of carbon-generated value changes. The valuation of carbon as constituted by and spread through markets and financial institutions is conversely a result of, among other things, how patterns of "the economy," such as consumption, financialization and neoliberalism, are built upon culturally shaped economic assumptions (see Best and Paterson 2010) and how these in turn depend upon specific social and cultural assumptions that human beings are rational actors who, as consumers, act in their best economic interest (see Sahlins 1976; Gausset, this volume). Many of the carbon-reducing initiatives we encounter thus build on attempts to secure the continuity of these cultural forms against the pressure of the growing climate crisis, whether this means "re-stabilizing" the narrative of capitalism as progress (Ervine 2012) or using imaginative "technical adjustments" to avoid questioning capitalism's social, political and economic relations (Günel 2019; Hougaard, this volume).

Here we return to the question of where changes come from. Early research into "the carbon economy" indicates that the way carbon is objectified as valuable is a result of capitalism as a cultural order or "system" (see Newell et al. 2012), but this research does not fully address how the system is in turn affected by the different cultures it encounters. Existing research on carbon often focusses on what carbon "does" in changing the values of individuals and collectivities in specific locations. But how is the value form itself transformed, and how is the dominant system of valuation transformed concurrently, if at all? For example, in the early days after the Kyoto Protocol and the Clean Development Mechanism, there were scandals such as when the production of hydrofluorocarbon (HFC) gasses in Chinese factories was enabled by selling carbon credits (e.g. MacKenzie 2009). In response, "the system" tried to correct itself, and subsequently, it has done so several times in the wake of resistance from, for example, critical journalists, scholars, buyers and indigenous groups. This has led to more elaborate verification mechanisms and standards that in turn have themselves become a set of commodities that "guarantee" the quality and value of carbon as a commodity (cf. Dalsgaard 2024). That is, despite the criticism, the deeper systemic capitalist structures and their approach to carbon have not changed.

Changes may develop from what in some contexts appear as new and external forces, as much as from existing (internal) dynamics. We will revisit this below, but first, it is necessary to stress that it would be naïve to separate these internal or external forces or dynamics into, on the one hand, carbon markets, scientists and experts, etc. as active (rational) agents behind systems and value forms that create changes through their governance of economic and political incentives, and, on the other hand, citizens, consumers or societies as passive recipients of such incentives for change that they may go along with or "resist," depending on their cultural predispositions. Such a separation would mean taking a superficial view of culture as made up of the elements that different collectivities or institutions "have" and can transfer to others, even if some meanings are "distorted" along the way. As mentioned already, culture can instead be seen as a relational and constantly developing set of positions and interpretations – as the meaningful distinctions people make in practice; distinctions that, to some degrees, are shared within social groups, corporations, networks or communities (cf. Sahlins 1985; Strathern 1995, 1997). In other words, if the value of carbon may be analysed as the result of human actors accepting or responding to external forces, this does not mean that the actors involved are passive and simply "receive" a value form. For example, carbon as a value form is rarely (and perhaps even "never") communicated solely in terms of a number (e.g. Lippert 2018), nor is it solely interpreted as such. Awareness of the climatic value of carbon, as much as how carbon works in economic and political institutions, has often been spread – albeit unevenly – throughout societies through communication via media and social networks. This communication has shaped people's cultural expectations around what is valuable about carbon and how (cf. Nerlich and Koteyko 2009; Brevini and Murdock 2015).

The contributions to this volume have something to say about this. For example, it is worth mentioning how the Dayak peatland dwellers in Central Kalimantan, Indonesia, have not grown carbon as a value form on their own. It has come to them through national peatland restoration projects promoted by government and international organizations in advance of any enumeration of their landscape's value as a carbon sink (Lounela, this volume). Also, the

Danish "carbon farmers" could be seen as responding to an externally promoted "configuration" of who and what they are when they physically spread carbon in the form of biochar on their fields (Hougaard, this volume). However, in both cases, a focus on this external origin would be to ignore the agency of the people in question as they react and respond in ways that stem from their own interests and prior experience. The peatland dwellers rely on their peatland channels for the sociality and material resources that can sustain their livelihoods. This impedes the wetland restoration needed to store carbon. The Danish carbon farmers are likewise interested in how new carbon forms can help them sustain a living. Yet this involves vying for a role in the Danish green transition that can be broader than just applying biochar to their fields. It also means choosing how to dispose of the farm-produced biomass that goes into making biochar in the first place, thus potentially expanding their agricultural production (and thus overall emissions) in the process. What the volume demonstrates is how the cultural reproduction of everyday life for Danish farmers, Indonesian peatland villagers and many others is, to various extents, transformed through carbon as a category of value, which leads to changing perceptions of forest, land, future and organizational forms. Speaking in terms of the changing semiotic meanings and conceptual value of carbon, carbon clearly takes on new value in all the contexts analysed in this volume. In some of these places, carbon has hardly been recognized on its own as a value form before, whereas in others it has been transformed from merely something that we breathe out to something that must be managed and contained. In other places, it may not be carbon as such that gains new value, but it is the land, the forest, the acts of consumption or the relations between people that are revalued and which have only recently become semiotically associated with carbon.

There is, in other words, nothing deterministic about the impact of external value forms (such as carbon) upon culture. Culture is an open-ended and unfinished repertoire that guides people's interpretations of value and life designs according to whatever aspirations any given individual or collectivity may pursue within the constraints of their surroundings. Sahlins emphasized how "different cultural orders have their own modes of historical action, consciousness, and determination – their own historical practice" (1985: 34). Cultural encounters between such orders would slowly transform while they interacted. While the culture of the neoliberal capitalist system (if it indeed is a system) may change at a different pace to that of communities that adopt carbon accounting or the commoditization of their natural resources into carbon credits, their various actions of consumption, resistance, adaptation, etc. nonetheless have an impact on what is sold and how the value of what is sold is established. As mentioned above, this gives rise to yet another set of "commodities" in the form of standards, methodologies, verifications and certifications that are necessary for carbon to be transformed into credits (cf. Dalsgaard 2024). These derive from the need to distinguish between that which can be counted as carbon and that which cannot; that which has a lot of climatic impact and that which has less of an impact. Where a standing forest – disregarding the multiple forms of life and social relationships it contains or sustains – may in one instance be regarded as merely a repository for storing carbon, it may under other circumstances become interpreted as a potential credit, which can be sold as soon as someone puts forward the threat of felling trees there. The system we refer to as capitalism may change slowly through its own "cultural circuit" of reflexivity (Thrift 2005), but the value forms sustaining capitalism may change at a quicker pace based upon the cultural encounters between marketized "objects" (carbon credits) and the different agents and networks of producers and consumers involved. If we identify such different tempi and degrees of change and discuss their discrepancies and differences from a perspective of value, we might be tempted to argue that there is a hierarchy of values at stake, where the making of an actual impact (and accepting the personal or collective "sacrifice" that this might entail) is subsumed to expectations of social status, or what one, with Pierre Bourdieu (1986), might discuss as capital within a given social field or group (see also Graeber 2013).

Emphasizing these inequalities does not merely point out that there are powerful systemic, infrastructural and global barriers opposing any deeper cultural change. It also questions how much a semiotic reference to one single element (carbon) can do to generate change. That "talk" about carbon can do *something* is certain. There are good examples of how specific actors start to see their own role and position in the world differently, and that some actors see nature and the climate differently. The use of the term Anthropocene may thus imply a new and different view of the relationship between humankind and the climate, even if the term is not focussed specifically on carbon emissions. Yet the awareness, the reflexivity and the internalization of everything pertaining to climate change, including the responsibility, agency and value associated with carbon as a prime agent in these changes, are still unevenly distributed, as is all knowledge (e.g. Barth 2002). To be sure, carbon – as a concept and a value form – has proven effective in raising awareness and reflexivity over climate justice and responsibility in some publics. Yet, its logic of enumeration has also been accused of turning justice into a matter of calculation rather than an engagement with the political and democratic "work of figuring out how to live in the world" (Beuret 2017: 15). Enumeration of carbon cannot help us solve the climate crisis on its own. That is, the actions and practices that the enumeration affects have clearly not yet been transformative enough, given that global emissions are currently still rising.

The chapters

The contributions to this volume engage carbon from a range of perspectives touching upon critical policy studies, social psychology, science and technology studies, sociology and anthropology. Accordingly, they do not present a uniform approach to "culture," and in this introduction, we have therefore provided a reading of how culture emerges or is implied across this variety of perspectives on carbon.

Siddharth Sareen and Bérénice Girard present us with a perspective that complicates the technocratic assumptions often entailed in the implementation of policy. They follow how an American model for the decarbonization of electricity supplied to energy-poor households travels to Europe. They emphasize the specific challenges to this process and how the model transforms and is adapted to local regulatory and demographic realities and infrastructures in the European countries where it is adopted (France, Spain and Italy). Carbon is often implicitly valued in such processes, and it lurks as a key value form driving the justification for making changes towards new infrastructures of electricity. Their chapter thus specifically shows how policy-driven change is never straightforward, even when societal majorities subscribe to decarbonization as a value in itself. Yet it seems that policies do create space for different emerging ideas of what counts as emission reductions.

Quentin Gausset's chapter asks what kind of change is generated by the deployment of carbon footprint calculators among Danish "eco-communities." Such calculators frequently align themselves with sociopsychological theories of behavioural change stemming from individual motivation. Gausset argues that these calculators embody sometimes problematic assumptions about what forms of consumption can help reduce emissions. In addition, the practical valuation and enumeration of emissions from a host of consumption practices perform a difficult balancing act between the accuracy of what counts and the flexibility that is needed to accommodate a wide range of uses and lifestyles. In Gausset's research, commitments in a social community – sometimes supported by calculators as technologies of quantification – prove to be stronger in motivating people to change by creating a space for support and reflexivity. This easily dispels the often-held technocratic assumption that solely providing individuals with data and information about emissions is sufficient for their reduction.

Inge-Merethe Hougaard has followed experiments that encourage the use of "biochar" fertilizer produced through pyrolysis. Danish farmers have been invited to apply biochar to their fields, which allegedly both binds carbon and enhances crop growth. Hougaard points out how an implicit and largely technically derived configuration of farmers as mere "users" of biochar ignores the wider practices of "carbon farming" that the production and distribution of biochar entails. For example, farmers are also expected to provide manure as biomass input to produce biochar, but biomass is also needed for the making of biogas. The latter has been crucial for emission reductions by replacing coal in the Danish energy sector. Finally, the application of biochar as a new emission-reducing practice provides yet another arena where carbon must be valued through counting and accounting. With multiple and unclear distinctions between "producers" and "users," potential disagreement emerges over the ownership of an emission reduction and towards whose account it gets credited.

Katinka Schyberg's chapter visits debates about emission reductions within the Danish Church. The Church is officially a state-governed institution – informally described as a "well-ordered anarchy" – bound by a strong tradition of independent and decentralized decision-making. This tradition, which has effectively kept the state from interfering in Church affairs, is challenged by carbon, which is a new phenomenon that a variety of Church actors now have to agree upon and situate both theoretically and practically within their organization. The chapter emphasizes how, on the one hand, efforts to calculate and display the Church's carbon footprint ended up uniting the Church against the odds and against further state intervention in terms of how the Church should engage in climate mitigation. On the other hand – and in seeming contrast to the first point – it also contributed to the preservation of both the decentralized structure of the Church and its ability to foster a heterogeneity of Christian ideas and political convictions.

Charlotte Bruckermann's chapter dives into the ways that carbon, in her case from China, is deployed to appropriate a diversity of cultural values. While carbon is introduced across China as a financially and economically quantified phenomenon, it is nonetheless challenged by the need to accommodate unforeseen externalities. As a result, carbon is technologically and bureaucratically coded with a broader "civilizing" force that stems from national ideologies of "ecological civilization." The tensions involved give rise to what Bruckermann refers to as relational frontlines between the economically oriented quantification and the qualitative values pursued by people in various positions. Here, Bruckermann centres on the case of rural afforestation workers, who do not uncritically accept a carbon-centred subjectivity when this is set against a state ideology anchored in a combination of neo-traditionalist and socialist values.

Anu Lounela takes the reader to Central Kalimantan, where Dayak villagers are engaged in damming the channels that crisscross the peatlands they inhabit. The purpose of this is to reverse the draining of the land both to reduce the risk of fires and to restore carbon in the landscape. The latter is expected to generate an income from the selling of carbon credits. However, the changes in the landscape are a source of local controversy because the raised water level impacts what can be grown in gardens and because the dams obstruct the channels' function as infrastructure for transportation and communication between dispersed villages and various sites for gardens, fishing, forestry or commercial timber production. Lounela argues that such restoration projects transform local social orders by commodifying the land as a "peat carbon frontier," and this "accumulation by restoration" generates conflicts between different value regimes as it impacts livelihoods and ownership of and access to land.

The chapter by Ingmar Lippert presents the reader with an experimental autoethnographic venture into an old lignite mining landscape in eastern Germany. The (previous) existence of carbon in the form of brown coal has transformed the land and the livelihoods of the local community through the forced movement (to some the destruction) of their village. Many traces remain of a troubled transformation, which is continually being re-interpreted in the light of changes in the present. This existence of the village in memories, in what Lippert refers to as "ont*political" designations, comes out through different visual and material representations of past events at an archive as well as at the site of the vanished village itself. The narrative of carbon's implied presence interchangeably challenges and justifies the extractivist commitments that lead to contemporary identifications with the landscape as one touched by the value of carbon in the specific form of lignite.

Andy Lautrup's Norwegian interlocutors are struggling over the right narrative for carbon's changing role in the future of Norwegian society. Carbon is tied into a national narrative of "goodness," because the extraction of oil has enabled the building of a strong Norwegian welfare state, yet issues of climate justice and responsibility for emissions are generating fault lines between different parts of the population. Lautrup portrays this as a problem of how to identify but also ideologically construct the proper "scale" for addressing carbon and the discussions it engenders. These discussions draw upon claims about the alleged exceptionality of the Norwegian oil industry and the "purity" of its oil, but also about how coupled or decoupled the producer of oil (and the oil itself as a substance) is from the responsibility for its combustion and the resulting emissions. Oil extraction is promoted by the industry as generating wealth that ensures the common good of the national community, but an activist counter-narrative emphasizes how the coupling of oil and emissions relates to a localized and historical responsibility to end oil and gas production to retain a liveable planet.

Altogether, the contributions thus point to several themes that address how carbon can be interpreted. First, carbon has a distinct *scale-making* potential as a physical resource due to it being an eternal yet metamorphic substance, but also socially it may transcend a diversity of scales and ideals. These could be organizational or political fields, like that of the Danish People's Church. The Church is officially a state-governed institution, but there are distinct challenges to the ways that carbon can be configured in relation to its organizational and material settings and traditions that emphasize decentralized decisions and independence from the state (Schyberg). Another example can be found in the intergenerational spaces where climate change is portrayed as a struggle between younger, globally oriented people – for example, in Norway – pushing for changes that established and middle-aged adults find difficult or too radical to realize because they see oil wealth as the foundation for the welfare state (Lautrup). Carbon as a lens and value form here raises questions about intergenerational as well as international responsibility and justice beyond the immediate spatiotemporal sites where oil extraction takes place.

Second, carbon has also become part of the renewed attachments that social communities develop to *landscapes* – both physical and informational – reconfigured through carbon extraction or sequestration (e.g. through reforestation or land restoration). The ways that carbon resources are produced, extracted or conversely stored – whether in mines, forests, peatlands or agricultural fields (Bruckermann; Hougaard; Lippert; Lounela) – is part of an ongoing transformation of people's relationships to the places where they live, work, play, trade, produce or consume. That carbon metrics become a central feature in territorialization or in place-making is not unique to rural settings or landscapes. It also appears in abstract value terms in areas re-imagined as pro-poor solar energy communities, but where top-down schemes meant to decarbonize energy consumption in these locations often require distinct adaptations to local geographies, demographics and legal realities (Sareen and Girard). Given the ubiquitous presence of carbon in life, land and energy forms, it is no surprise that it enters as well as transcends a diversity of cultural scales and relationships in both time and space.

Third, this capacity for ubiquity and transcendence can present challenges when one tries to understand the consequences of choosing one way of living with carbon over another – for those who indeed feel they have a choice – or for delivering the desired and frequently promised "ease" by which (especially) political or corporate actors imagine the transitions to societies with low emission livelihoods. At a more overarching level, such discussions touch upon the multiple *imaginations*, realities and even alienations that carbon sustains and changes (see Norgaard 2018) and how understandings of carbon are part of and informed by a diversity of calculative logics, quantifications, accounting measures and systems, which people adopt or adapt to in a variety of ways depending on factors such as sociocultural context or personal interest (e.g. Bruckermann; Gausset; Hougaard; Sareen and Girard). The disciplinary approaches of the natural sciences and economics have had a huge impact on how different individuals and collectivities identify carbon in different spheres of life, but the contributions here show through qualitative and mostly ethnographic studies that what we might refer to as the "recipients" of the numerical categories of carbon are actually configured in a variety of ways. For example, some small communities actively make sense of their own carbon footprints to change behaviours and lifestyles as "consumers" of carbon (Gausset), whereas others feel that they are configured more passively as mere "users" of carbon – here in the form of biochar – as a new resource that combines agricultural fertilization with decarbonization (Hougaard). The enumerative value of carbon does not determine nor does it always preclude a specific social level of engagement or agency.

Fourth, the different contributions thus together paint a picture of carbon as a cultural object with *a multiplicity of meanings* across scales, landscapes and imaginations. Such multiplicities are revealed through the contributors' pursuit of qualitative approaches that recognize the irreducible character of the cultural complexities that carbon gets entangled with. Such complexities can best be addressed through careful attention to how meanings are ascribed to carbon, what affects these meanings and how they shift under a variety of circumstances and contexts. The varied conceptual existences of carbon outlined in this volume are driven by social and culturally informed interpretations as much as they are sources for social and cultural life in turn (cf. Strathern 1997: 44). Having said that, we might sum up all of the above with the following overarching questions: how is carbon involved in cultural change? And how does carbon, as a new cultural artefact, change existing norms and practices? In his contribution, Gausset addresses change directly as "behavioural change," but change in a more general form is present in the other chapters too, in terms of relating to changing landscapes of production or extraction (Hougaard, Lippert and Lounela) or in discussing how changing identities relate to the resources that sustain one's society (Hougaard and Lautrup) or one's community, whether material or spiritual (Gausset and Schyberg). The chapters also relate to changing energy infrastructures (Sareen) and to the forms of production or extraction related to carbon-intensive sectors such as agriculture or afforestation, processes which nonetheless are often still managed in ways that impinge upon everyday social lives (Hougaard and Bruckermann). Finally, a large part of the foreseen changes revolve around imaginaries and imaginations of different spatiotemporal scales (Lautrup and Bruckermann). Setting a clear vision for what kind of society is to be achieved within the parameters set out by, for example, the agreed-upon reductions specified in the 2015 Paris Agreement is frequently a starting point for the pursuit of change, but it also points to the tensions between visions of the change desired, ideas of how to realize this through interventions and what the practical effects of the intervening actions might be.

Concluding remarks

This introduction and the subsequent chapters demonstrate how important it is to see carbon as part of a cultural transformation, and in aggregate the contributions point to the diversity that a conception of culture change can cover. This may include a multiplicity of aspects such as changing behaviour, consumption and living (Gausset); inscribing memory into archives or landscapes (Lippert); the changing relations between land and livelihood (Lounela); the organizing of Church values (Schyberg); new configurations of professional identities (Hougaard); how a national identity may be facing change (Lautrup); the difference between (national state) ideology and (local) practice (Bruckermann); and the contrast between models and implementation in the roll-out of specific sociotechnical plans for the provision of low-cost and low-carbon electricity to energy-poor households (Sareen and Girard). To some, it may come as a surprise that technological interventions based on a culturally vague entity such as "carbon" come out as ambiguous as they do (see Sareen and Girard), but such diversity should not come as a surprise to a keen cultural analyst, especially not given the geographical spread of the contributions and the different circumstances at stake in each case.

The majority of the contributions engage in an ethnographic dialogue with other perspectives on how carbon appears (or disappears) in different contexts of social or cultural affairs. Retaining an ethnographic and qualitative perspective on what carbon "means" in different contexts is crucial if we are to understand the diverse outcomes of different policies (technocratic and bureaucratic) and transformations. Naturally, this perspective cannot stand alone.

No perspective can. However, to understand how the climate crisis presents an urgency, which to many societies appears unprecedented, this perspective is crucial, even if the sociocultural changes that may stem from climate change are still unknown. We can say for certain that carbon will remain a valuable element for many human cultures in the foreseeable future, but the ways in which it will transform or be transformed by these remains to be seen.

Acknowledgements

This introduction owes a great deal to the co-editors of the volume, Andy Lautrup, Katinka Schyberg and Ingmar Lippert, and its formulations have been vastly improved after Jeremy Gunson's careful reading. The final responsibility for its formulations, of course, remains mine.

Notes

1 In a volume discussing the politics of green transformations, Ian Scoones, Peter Newell and Melissa Leach (2015) point to four broad narratives of such transformations. Each narrative presents its own emphasis on and framing of both the problem and the solution. They are thus political in that they provide different perspectives on what is to be changed, and how this change is to occur. This is clear in how Scoones, Newell and Leach identify the narratives – technocentric, marketized, state-led and citizen-led – based upon who and what is ascribed the primary role and agency within each of them.
2 Compliance markets are regulated and aimed at helping industry achieve government-set targets for emission reductions, while voluntary markets are typically used by industry or other organizations to purchase emission reductions based on their own motivation.
3 Reduced Emissions from Deforestation and Forest Degradation – the + refers to the sustainable management, conservation and enhancement of forest carbon stocks (e.g. Paladino and Fiske 2017).
4 Emphasizing a symbolic and semiotic ordering of reality does not mean that materiality is a mere appendix to an instrumental or deterministic idea of culture. Although with varying emphasis on the semiotic and symbolic versus the affordances of materiality, both Sahlins (1985) and Graeber (2001) agree there is a practical constitution and reconstitution of value taking place in human speech and action.

References

Ahonen, Hanna-Mari, Juliana Kessler, Axel Michaelowa, Aglaja Espelage and Stephan Hoch. 2022. "Governance of Fragmented Compliance and Voluntary Carbon Markets Under the Paris Agreement." *Politics and Governance* 10 (1): 235–45.

Albert, Osei-Owusu Kwame, Marianne Thomsen, Jonathan Lindahl, Nino Javakhishvilie Larsen and Caro Dario. 2020. "Tracking the Carbon Emissions of Denmark's Five Regions from a Producer and Consumer Perspective." *Ecological Economics* 177: 106778.

Appel, Hannah, Arthur Mason and Michael Watts (eds.). 2015. *Subterranean Estates*. Ithaca: Cornell University Press.

Barth, Fredrik. 2002. "An Anthropology of Knowledge." *Current Anthropology* 43 (1): 1–18.
Best, Jacqueline and Matthew Paterson. 2010. "Introduction: Understanding Cultural Political Economy." In Cultural Political Economy, edited by Jaqueline Best and Matthew Paterso, 1–25. New York: Routledge.
Beuret, Nicholas. 2017. "Counting Carbon: Calculative Activism and Slippery Infrastructure." *Antipode* 49 (5): 1164–85. Online early view: https://doi.org/10.1111/anti.12317
Bigger, Patrick. 2016. "Regulating Fairness in the Design of California's Cap-and-Trade Market." In *The Carbon Fix*, edited by Stephanie Paladino and Shirley J. Fiske, 119–34. New York: Routledge.
Blühdorn, Ingolfur. 2007. "Sustaining the Unsustainable: Symbolic Politics and the Politics of Simulation." *Environmental Politics* 16 (2): 251–75. https://doi.org/10.1080/09644010701211759
Bourdieu, Pierre. 1986. "The Forms of Capital." In *Handbook of Theory and Research for the Sociology of Education*, edited by John G. Richardson, 241–58. New York: Greenwood Press.
Bowen, Frances and Bettina Wittneben. 2011. "Carbon Accounting: Negotiating Accuracy, Consistency and Certainty across Organisational Fields." *Accounting, Auditing & Accountability Journal* 24 (8): 1022–36. https://doi.org/10.1108/09513571111184742
Boyer, Dominic. 2019. *Energopolitics: Wind and Power in the Anthropocene*. Durham: Duke University Press.
Brevini, Benedetta and Graham Murdock (eds.). 2015. *Carbon Capitalism and Communication*. London: Palgrave Macmillan.
Bridge, Gavin. 2010. "Resource Geographies I: Making Carbon Economies, Old and New." *Progress in Human Geography* 35 (6): 820–34. https://doi.org/10.1177/0309132510385524
Brightman, Robert. 1995. "Forget Culture: Replacement, Transcendence, Relexification." *Cultural Anthropology* 10 (4): 509–46. https://doi.org/10.1525/can.1995.10.4.02a00030
Buck, Holly Jean. 2022. "Mining the Air: Political Ecologies of the Circular Carbon Economy." *Environment and Planning E: Nature and Space* 5 (3): 1086–105. https://doi.org/10.1177/25148486211061452
Bulkeley, Harriet and Peter Newell. 2010. *Governing Climate Change*. London: Routledge.
Bulkeley, Harriet, Matthew Paterson and Johannes Stripple (eds.). 2016. *Towards a Cultural Politics of Climate Change*. Cambridge: Cambridge University Press.
Corbera, Esteve and Heike Schroeder. 2011. "Governing and Implementing REDD+." *Environmental Science & Policy* 14: 89–99. https://doi.org/10.1016/j.envsci.2010.11.002
Dalsgaard, Steffen. 2013. "The Commensurability of Carbon." *HAU: Journal of Ethnographic Theory* 3 (1): 80–98. https://doi.org/10.14318/hau3.1.006
Dalsgaard, Steffen. 2022. "Tales of Carbon Offsets: Between Experiments and Indulgences?" *Journal of Cultural Economy* 15 (1): 52–66. https://doi.org/10.1080/17530350.2021.1977675
Dalsgaard, Steffen. 2024. "The Controversy of Voluntary Carbon Offsetting: Compliance by Proxy." In *The Anthropology of Compliance*, edited by Will Rollason and Eric Hirsch, 128–50. Oxford: Berghahn.
Di Muzio, Tim. 2015. *Carbon Capitalism*. Lanham, MD: Rowman & Littlefield Publishers.
Ehrenstein, Vera and Fabian Muniesa. 2013. "The Conditional Sink: Counterfactual Display in the Valuation of a Carbon Offsetting Reforestation Project." *Valuation Studies* 1 (2): 161–88. https://doi.org/10.3384/vs.2001-5992.1312161
Ervine, Kate. 2012. "The Politics and Practice of Carbon Offsetting: Silencing Dissent." *New Political Science* 34 (1): 1–20. https://doi.org/10.1080/07393148.2012.646017

Ervine, Kate. 2018. *Carbon*. Cambridge: Polity.
Gifford, Lauren. 2020. "'You Can't Value What You Can't Measure': A Critical Look at Forest Carbon Accounting." *Climatic Change* 161: 291–306. https://doi.org/10.1007/s10584-020-02653-1
Girvan, Anita. 2018. *Carbon Footprints as Cultural-Ecological Metaphors*. New York: Routledge.
Goodman, Michael and Emily Boyd. 2011. "A Social Life for Carbon? Commodification, Markets and Care." *The Geographical Journal* 177 (2): 102–09. https://doi.org/10.1111/j.1475-4959.2011.00401.x
Graeber, David. 2001. *Toward an Anthropological Theory of Value*. London: Palgrave Macmillan.
Graeber, David. 2013. "Postscript: It Is Value that Brings Universes into Being." *HAU: Journal of Ethnographic Theory* 3 (2): 219–43. https://doi.org/10.14318/hau3.2.012
Greenleaf, Maron. 2020. "The Value of the Untenured Forest: Land Rights, Green Labor, and Forest Carbon in the Brazilian Amazon." *Journal of Peasant Studies* 47 (2): 286–305. https://doi.org/10.1080/03066150.2019.1579197
Grubb, Michael. 2003. "The Economics of the Kyoto Protocol." *World Economics* 4 (3): 143–89.
Gupta, Aarti, Eva Lövbrand, Esther Turnhout and Marjanneke Vijge. 2012. "In Pursuit of Carbon Accountability: The Politics of REDD+ Measuring, Reporting and Verification Systems." *Current Opinion in Environmental Sustainability* 4 (6): 726–31. https://doi.org/10.1016/j.cosust.2012.10.004
Günel, Gökce. 2016. "What Is Carbon Dioxide?" *PoLAR* 39 (1): 33–45. https://doi.org/10.1111/plar.12129
Günel, Gökce. 2019. *Spaceship in the Desert*. Durham: Duke University Press.
Helm, Dieter and Cameron Hepburn (eds.). 2009. *The Economics and Politics of Climate Change*. Oxford: Oxford University Press.
Horton, David. 2003. "Green Distinctions: The Performance of Identity among Environmental Activists." *The Sociological Review* 51 (s2): 63–77.
Howe, Cymene. 2019. *Ecologics: Wind and Power in the Anthropocene*. Durham: Duke University Press.
Hughes, David D. 2013. "Climate Change and the Victim Slot: From Oil to Innocence." *American Anthropologist* 115 (4): 570–81. https://doi.org/10.1111/aman.12044
Klein, Naomi. 2014. *This Changes Everything*. New York: Simon & Schuster.
Knox, Hannah. 2020. *Thinking Like a Climate*. Durham: Duke University Press.
Knox-Hayes, Janelle. 2013. "The Spatial and Temporal Dynamics of Value in Financialization: Analysis of the Infrastructure of Carbon Markets." *Geoforum* 50: 117–28. https://doi.org/10.1016/j.geoforum.2013.08.012
Knox-Hayes, Janelle. 2016. *The Cultures of Markets*. Oxford: Oxford University Press.
Leach, Melissa and Ian Scoones (eds.). 2015. *Carbon Conflicts and Forest Landscapes in Africa*. London: Routledge. doi.org/10.4324/9781315740416
Lindman, Åsa, Kristina Ek and Patrik Söderholm. 2013. "Voluntary Citizen Participation in Carbon Allowance Markets." *Climate Policy* 13 (6): 680–97. https://doi.org/10.1080/14693062.2013.810436
Lippert, Ingmar. 2013. *Enacting Environments: An Ethnography of the Digitalisation and Naturalisation of Emissions*. PhD thesis, University of Augsburg.
Lippert, Ingmar. 2015. "Environment as Datascape. Enacting Emission Realities in Corporate Carbon Accounting." *Geoforum* 66: 126–35.
Lippert, Ingmar. 2018. "On Not Muddling Lunches and Flights." *Science & Technology Studies* 31 (4): 52–74. https://doi.org/10.23987/sts.66209
Lohmann, Larry. 2010. "Uncertainty Markets and Carbon Markets: Variations on Polanyian Themes." *New Political Economy* 15 (2): 225–54. https://doi.org/10.1080/13563460903290946

Lohmann, Larry. 2011. "The Endless Algebra of Carbon Markets." *Capitalism Nature Socialism* 22 (4): 93–116. https://doi.org/10.1080/10455752.2011.617507

Lövbrand, Eva and Johannes Stripple. 2011. "Making Climate Change Governable: Accounting for Carbon as Sinks, Credits and Personal Budgets." *Critical Policy Studies* 5 (2): 187–200. https://doi.org/10.1080/19460171.2011.576531

Lovell, Heather, Harriet Bulkeley and Diana Liverman. 2009. "Carbon Offsetting: Sustaining Consumption?" *Environment and Planning A* 41: 2357–79. https://doi.org/10.1068/a40345

Lovell, Heather and Diana Liverman. 2010. "Understanding Carbon Offset Technologies." *New Political Economy* 15(2):255–73. https://doi.org/10.1080/13563460903548699

Lovell, Heather and Donald MacKenzie. 2012. "Accounting for Carbon: The Role of Accounting Professional Organisations in Governing Climate Change." In *The New Carbon Economy*, edited by Boykoff Newell and Boyd. Oxford: Wiley-Blackwell.

Lovell, Heather. 2015. *The Making of Low Carbon Economies*. London: Routledge.

MacKenzie, Donald. 2009. "Making Things the Same: Gases, Emission Rights and the Politics of Carbon Markets." *Accounting, Organizations and Society* 34: 440–55. https://doi.org/10.1016/j.aos.2008.02.004

Mahanty, S., Sarah Milne, Wolfram Dressler and Colin Filer. 2012. "The Social Life of Forest Carbon." *Human Ecology* 40 (5): 661–64. https://doi.org/10.1007/s10745-012-9524-1

Marres, Noortje. 2011. "The Costs of Public Involvement: Everyday Devices of Carbon Accounting and the Materialization of Participation." *Economy and Society* 40 (4): 510–33. https://doi.org/10.1080/03085147.2011.602294

Mitchell, Timothy. 2011. *Carbon Democracy*. London: Verso.

Moser, Stephanie and Silke Kleinhückelkotten. 2018. "Good Intents, but Low Impacts: Diverging Importance of Motivational and Socioeconomic Determinants Explaining Pro-Environmental Behavior, Energy Use, and Carbon Footprint." *Environment and Behavior* 50 (6): 626–56.

Murdiyaso, Daniel, Maria Brockhaus, William Sunderlin and Lou Verchot. 2012. "Some Lessons Learned from the First Generation of REDD+ Activities." *Current Opinion in Environmental Sustainability* 4: 678–85. https://doi.org/10.1016/j.cosust.2012.10.014

Nerlich, Brigitte and Nelya Koteyko. 2009. "Compounds, Creativity and Complexity in Climate Change Communication. The Case of 'Carbon Indulgences'". *Global Environmental Politics* 19 (3): 345–53. https://doi.org/10.1016/j.gloenvcha.2009.03.001

Newell, Peter and Matthew Paterson. 2010. *Climate Capitalism*. Cambridge: Cambridge University Press.

Newell, Peter, Maxwell Boykoff and Emily Boyd (eds.). 2012. *The New Carbon Economy*. Oxford: Wiley-Blackwell.

Norgaard, Kari. 2011. *Living in Denial: Climate Change, Emotions, and Everyday Life*. Cambridge, Mass.: MIT Press.

Norgaard, Kari. 2018. "The Sociological Imagination in a Time of Climate Change." *Global and Planetary Change* 163: 171–76. https://doi.org/10.1016/j.gloplacha.2017.09.018

Otto, Ton. 2024. "Anthropological Perspectives on Value and Valuation." In *The Routledge International Handbook of Valuation and Society*, edited by Anne Krüger, Thorsten Peetz and Hilmar Schaefer, 23–33. New York: Routledge.

Otto, Ton and Rane Willerslev. 2013. "Value as Theory." *Special issue of HAU: Journal of Ethnographic Theory* 3: 1–2. https://doi.org/10.14318/hau3.1.002

Paladino, Stephanie and Shirley Fiske (eds.). 2017. *The Carbon Fix*. London: Routledge.

Pascoe, Sophie. 2018. "Interrogating Scale in the REDD+ Assemblage in Papua New Guinea." *Geoforum* 96: 87–96. https://doi.org/10.1016/j.geoforum.2018.08.007

Paterson, Matthew and Johannes Stripple. 2010. "My Space: Governing Individuals' Carbon Emissions." *Environment and Planning D: Society and Space* 28: 341–62. https://doi.org/10.1068/d4109

Paterson, Matthew and Johannes Stripple. 2012. "Virtuous Carbon." *Environmental Politics* 21 (4): 563–82. https://doi.org/10.1080/09644016.2012.688354

Pearse, Rebecca. 2018. *Pricing Carbon in Australia*. London: Routledge.

Rice, Jennifer. 2010. "Climate, Carbon, and Territory: Greenhouse Gas Mitigation in Seattle, Washington." *Annals of the Association of American Geographers* 100 (4): 929–37. https://doi.org/10.1080/00045608.2010.502434

Roncoli, Carla, Todd Crane and Ben Orlove. 2009. "Fielding Climate Change in Cultural Anthropology." In *Anthropology and Climate Change*, edited by Susan Crate and Mark Nuttall, 87–115. Walnut Creek: Left Coast Press.

Sahlins, Marshall. 1976. *Culture and Practical Reason*. Chicago: The University of Chicago Press.

Sahlins, Marshall. 1985. *Islands of History*. Chicago: The University of Chicago Press.

Sahlins, Marshall. 1999. "Two or Three Things I Know About Culture." *Journal of the Royal Anthropological Institute (N.S.)* 5 (3): 399–421. https://doi.org/10.2307/2661275

Sareen, Siddharth and Katja Müller (eds.). 2023. *Digitisation and Low-Carbon Energy Transitions*. Cham: Palgrave Macmillan.

Scoones, Ian, Peter Newell and Melissa Leach. 2015. "The Politics of Green Transformations." In *The Politics of Green Transformations*, edited by Melissa Leach, Ian Scoones and Peter Newell, 1–24. London: Routledge. https://doi.org/10.4324/9781315747378

Singer, Merrill, Jose Hasemann and Abigail Raynor. 2016. "'I Feel Suffocated:' Understandings of Climate Change in an Inner City Heat Island." *Medical Anthropology* 35 (6): 453–63. https://doi.org/10.1080/01459740.2016.1204543

Strathern, Marilyn. 1995. "The Nice Thing About Culture Is That Everyone Has It." In *Shifting Contexts*, edited by Marilyn Strathern, 153–76. London: Routledge.

Strathern, Marilyn. 1997. "The Work of Culture: An Anthropological Perspective." In *Culture, Kinship and Genes*, edited by Angus Clarke and Evelyn Parsons, 40–53. Palgrave Macmillan.

Thrift, Nigel. 2005. *Knowing Capitalism*. London: SAGE.

Underwood, Anthony and Anders Fremstad. 2018. "Does Sharing Backfire? A Decomposition of Household and Urban Economies in CO2 Emissions." *Energy Policy* 123: 404–13. https://doi.org/10.1016/j.enpol.2018.09.012

Walenta, Jayme. 2021. "The Making of the Corporate Carbon Footprint: The Politics Behind Emission Scoping." *Journal of Cultural Economy* 14 (5): 533–48. https://doi.org/10.1080/17530350.2021.1935297

Wang, Yiting and Catherine Corson. 2015. "The Making of a 'Charismatic' Carbon Credit: Clean Cookstoves and 'Uncooperative' Women in Western Kenya." *Environment and Planning A* 47: 2064–79. https://doi.org/10.1068/a130233p

Wettestad, Jørgen and Lars Guldbrandsen (eds.). 2018. *The Evolution of Carbon Markets*. London: Routledge.

Whitington, Jerome. 2016. "Carbon as a Metric of the Human." *PoLAR* 39 (1): 46–63. https://doi.org/10.1111/plar.12130

Whitmarsh, Lorraine and Saffron O'Neill. 2010. "Green Identity, Green Living? The Role of Pro-environmental Self-identity in Determining Consistency across Diverse Pro-environmental Behaviours." *Journal of Environmental Psychology* 30: 305–14. https://doi.org/10.1016/j.jenvp.2010.01.003

Yamin, Farhana (ed.). 2005. *Climate Change and Carbon Markets*. London: Routledge. doi.org/10.4324/9781849770781

Yocum, Heather. 2016. "'It *Becomes* Scientific…:' Carbon Accounting for REDD+ in Malawi." *Human Ecology* 44: 677–85. https://doi.org/10.1007/s10745-016-9869-y

1 Renewable energy communities for energy-poor households

Policy mobility challenges in urban and peri-urban European contexts

Siddharth Sareen and Bérénice Girard

Introduction

The proposition of carbon emission reduction is simultaneously both personal and impersonal. It becomes personal when mobilized in particular conversations and measurements, such as those of individual aviation-related emissions or emissions related to using an automobile for everyday transport. It is also personal when pertaining to one's source of fuel to heat or cool a home, directly (like a wood oven) or indirectly (like an air conditioner run on coal-powered or hydro-powered electricity), or to cook (on a gas or induction stove). These discussions and metrics tend to be linked to arguments for behavioural change and ethics (Hedberg 2018) and arguably distract focus from the systemic nature of greenhouse gas emissions linked to larger structures that shape individual energy practices (Capstick et al. 2014). Debates about individual energy use often get contentious because they can easily pertain to excessive or elite consumption in an unequal world (Nielsen et al. 2021). This can elicit defensive responses from powerful beneficiaries of the status quo or lead to "victim-blaming" or putting the burden of change on those who already consume energy modestly and may even be exposed to energy poverty[1] (Sardo 2023), which runs the risk of an antipoor bias in environmentalism. Thus, the personalization of carbon emission reduction presents a difficult route to consensus and rapid action. Yet when carbon emission reduction is treated impersonally, it essentially becomes an artefact buried in documents too technocratic to be anyone's concern beyond experts and decision-makers (Chilvers and Longhurst 2016).

So, too prone to argumentative inertia when personal and too liable to be left alone in a corner as a concern relegated to experts when impersonal – is there a third way to make sense of carbon emission reduction that can precipitate action? In this chapter, we argue that there is, and suggest that the key to doing so lies in recognizing the sociocultural constitution of carbon emission reduction. It is in the embrace of sociocultural dynamics that carbon finds meaning and can plausibly find the desired movement towards climate change mitigation. We suggest that carbon can be analysed as socioculturally constitutive of practices that seek to govern and mitigate climate change, and ontologically distinguish the sociocultural dynamics – too often subsumed under

DOI: 10.4324/9781003478669-2

"aspects" – that mitigation projects face from carbon itself. In this chapter, we study sociocultural dynamics as they are envisioned by actors implementing a socially inclusive low-carbon energy project within diverse institutional configurations. We analyse how these visions interact with the context-specific opportunities and constraints that these actors face and how they evolve during policy definition and implementation. This chimes well with the relational aspects of carbon and cultural change explicated in the introduction to this volume.

It would be facile to posit the sociocultural dimension of carbon emission reduction as some sort of uncontested fact. In fact, this claim is far from commonly acknowledged, and it is deeply countered by those who argue that carbon emission reduction is primarily a techno-economic challenge. Techno-economic analysis still exerts a powerful hold over the human imagination when it comes to climate change mitigation solutions (Birch 2017). This necessitates efforts to elucidate the sociocultural dynamics related to carbon, a task which this book undertakes. In this chapter, we show how an important part of understanding the sociocultural embeddedness of carbon emission reduction is revealed in the messy details of how these mitigation solutions are implemented. To draw out how sociocultural aspects are taken into account by policymakers in the Eurosolar4All project, we draw on the literature on policy mobility (see, e.g., Cochrane and Ward 2012). This attention to the lived realities of policymakers resonates with the classical call to studying up in anthropology (Nader 1972).

The original project at the heart of the effort this chapter unpacks was a quintessentially techno-economic undertaking. It was piloted in the late 2010s by the New York State Energy Research and Development Authority (NYSERDA). For NYSERDA, combining energy poverty alleviation with renewable energy promotion at the community-scale served a double objective. First, it promoted the adoption of small-scale solar energy beyond upper-middle-class households who could afford to invest in solar modules. By investing in community solar plants and using virtual net metering to offer consumption of this low-carbon energy source at a low tariff for low-income households, public authorities could engender a pro-poor solar energy rollout and in turn create more demand for decentralized solar energy. Second, by showing that a large range of households could benefit from decentralized renewable energy sources, public authorities could accelerate public support for and adoption of these sources. This in turn helped towards reaching ambitious climate mitigation targets more quickly and at lower cost, as these community solar plants could be replicated and easily scaled up to multiply beneficiaries and thus solar energy users.

Yet, as we will show, the techno-economic results cannot be separated from the sociocultural enablement of this technology for carbon emission reduction. The success of the pilot scheme inspired NYSERDA to expand this initiative state-wide to the tune of 820 megawatts (MW), corresponding to approximately a billion dollars of investment in community-scale solar plants. It also inspired a dozen organizations in Europe to cooperate and emulate this approach in different contexts with diverse legislative and sociocultural conditions. This led to the Eurosolar4All project, funded within the Horizon 2020 framework programme of the European Commission from October 2021 to September 2024.[2] While other chapters in this volume use ethnographic methods to study the sociocultural dimensions of carbon, we employ a complementary approach. This chapter draws on our observations from the first two years of Eurosolar4All's efforts through our experience of conducting social impact assessments of these interventions in four European contexts. As such, we were able to observe the debates, conversations and hurdles faced by a group of actors as they set upon designing and implementing a low-carbon energy project. We are therefore able to identify where sociocultural complexities were considered by these actors and where they may have been ignored or underestimated in the design and implementation processes. We do not report the content of the social impact assessments, nor do we offer fine-grained empirical details of each pilot, as these matters are reserved for other outputs stemming from the project. Rather, our concern here is to analyse how the very process of transposing this scheme from the United States to four diverse European contexts presents a window into the sociocultural constitution of carbon emission reduction schemes.

Towards this, we next review the literature on the intersection of renewable energy communities (RECs) and energy poverty in Europe. The main claim that emerges here is that there is a synergistic relationship between promoting community-scale renewable energy projects and schemes for energy poverty alleviation. This situates the Eurosolar4All project within scholarship that argues for policy orientation and experimentation along such lines. Thereafter, we briefly introduce the analytical framework of policy mobility and present our methods; we employ this analytical framework primarily to structure our argument. Then we describe and analyse the policy transposition process with an empirical focus on and across the four comparative case studies. This entails policy mobility from Solar for All to the Eurosolar4All project, where the original project was adapted to fit within the logic of a Horizon 2020 project. It then entails another policy mobility from the ideal-type envisaged in the Eurosolar4All project plans to its messy implementation in each of the four contexts – from large cities (Barcelona in Spain and Rome in Italy) to a smaller municipality (Almada in Portugal) and a peri-urban federation of small municipalities (Coeur de Savoie in France). Thereafter, we offer a discussion and a conclusion. We emphasize the key takeaways in terms of the sociocultural characteristics of carbon emission reduction that emerge from our study and reflect upon their implications for a synergistic agenda for RECs and energy policy alleviation both in and beyond Europe.

Renewable energy communities and energy poverty alleviation in Europe

In the late 2010s, two important directives emerged at the European level on RECs. The first was the Revised Renewable Energy Directive, or RED II,

EU/2018/2001 (European Union 2018); the second, the Directive on Common Rules for the Internal Electricity Market, EU/2019/944 (European Union 2019). Two related initiatives were also launched: the Energy Communities Repository and the Rural Energy Community Advisory Hub. Based on these regulations and initiatives, as well as on relevant literature, here we define RECs as legal entities which are based on open and voluntary participation, focus on renewable energy, are often partly community or citizen-owned and controlled by their shareholders and/or by residents, and prioritize economic, social and environmental benefits to the community before financial profits (Bielig et al. 2022). Energy communities are supposed to be organized collectively and to serve as a vehicle for citizen-led transformation. They allow the involvement of "community members in decision-making about energy production, consumption and the distribution of associated benefits" (Bielig et al. 2022, p.1).

Moreover, according to RED II, RECs are supposed to foster energy justice, energy democracy and social acceptability. They are especially envisioned as a resource to enable energy poverty alleviation. Thus, the RED II directive (European Union 2018, §67) states that:

> Empowering jointly acting renewables self-consumers also provides opportunities for renewable energy communities to advance energy efficiency at household level and helps fight energy poverty through reduced consumption and lower supply tariffs. Member States should take appropriate advantage of that opportunity by, inter alia, assessing the possibility to enable participation by households that might otherwise not be able to participate, including vulnerable consumers and tenants.

This is a sophisticated vision of carbon emission reduction, not simply related to climate mitigation targets, but one that actively seeks to incorporate a social goal in the manner and ontology of carbon emission reduction itself. It is also striking in its ambition for a sector historically characterized by a lack of and barriers to broad public engagement, given the technocratic and top-down tendencies of energy governance (MacArthur 2016). Concomitantly, such a formal link between carbon emission reduction and energy poverty alleviation in European level policy necessitates an appreciation of carbon as socioculturally constituted and carbon emission reduction as entangled in sociocultural processes. These policies are translated into national level effects through periodically revised National Energy and Climate Plans of European Union Member States, including the four in our study.

Crucially, then, as Hanke et al. (2021, p.2) underline, RED II does not provide "details on how to achieve RECs' social role in practice." Clearly, RECs have a high potential to alleviate energy poverty, which we understand, in line with Bouzarovski and Petrova (2015, p.31), as "the inability to attain a socially and materially necessitated level of domestic energy services." RECs can provide access to lower energy tariffs and additional sources of income and can have potential positive impacts on energy behaviour and low-income household economies (Roberts et al. 2022). However, research shows that they are frequently composed of relatively privileged members with economic capital and technical know-how, whereas vulnerable households often lack the networks, financial resources or time to join (Hanke et al. 2021). Moreover, RECs often do not reach and engage with energy-vulnerable households, do not have specific procedures to include them, or do not have the time or willingness to develop specific engagement activities or collaborate with local actors who might help develop these activities (Hanke et al. 2021). These attributes must be proactively designed and built into RECs to avoid exacerbating injustices (Hoicka et al. 2021; Sareen et al. 2023).

The literature on energy poverty alleviation has grown substantially over the past decade (Jiglau et al. 2023) and will doubtless expand to take on board the implications of RECs in the course of the 2020s. This chapter is not the place for an elaborate review of this literature, but suffice it to say that without pre-emptive efforts to safeguard against tendencies for inequity, exclusion and co-optation by elite and incumbent actors, many of the trends apparent in scholarship on energy poverty are likely to be reproduced in the rollout of RECs. Community in itself is no cure-all (Aiken 2017). Yet, given policy traction for these two trends of distributed renewable energy rollout and energy poverty alleviation to converge, and their virtuous potential towards a desirable end, enthusiasm by researchers and policymakers alike is understandable, and worthy of an embrace, albeit a critical one.

Inspired by this literature, but also circumspect about possible headway both in theory and in practice, we propose a change of focal point: rather than focus on RECs' engagement with vulnerable households, we investigate how RECs directly targeted at vulnerable households are designed and implemented in diverse contexts. How was the sociocultural context of carbon emission reduction considered by the different actors in project design and implementation? In particular, how did each intervention engage with energy poverty alleviation? How did these interventions – prompted by the need for carbon emission reduction while also recognizing the need to address other societal challenges like inequity and exclusion – balance the distribution of material benefits, the establishment of RECs through citizen engagement and the empowerment of vulnerable households?

Analytical framework and methods: operationalizing a dual policy mobility lens

Since the 2010s, policy mobility scholarship has taken on a distinct urban focus. Prior to this, the literature focussed on the history of transnational municipal movements (Clarke 2012). The urban focus stems from a realization within human geography and urban studies scholarship that cities are often the sites of the exchange and assembly of particular policies, flows and configurations that in themselves are profoundly geographical subject matters (Cochrane and Ward 2012). Examining processes of policy mobility allows us to underline the importance of place and context in the enactment of any given intervention, as its embedding and operationalization within a context in turn acts back upon and reframes it. Thus, the understanding of policy mobility that we mobilize here concerns a two-way process: first, how a policy from elsewhere is brought into a context by particular actors, and second, how it is shaped through the work of situated operationalization. In more abstract terms, Prince (2017) has articulated policy mobility as comprising the work of assemblage that is territorialized in territories with different topologies, thus producing different ideas of the global and the local through general visions and specific manifestations. Hence, a scheme like Solar for All, when mobilized through a project like Eurosolar4All, takes on particular shapes and forms in the four pilot project contexts. These shapes and forms are co-produced by the assemblage of actors engaging with the project, both in its journey from an idea-from-elsewhere to specific contexts and in the concrete manifestations of that idea in the process of local implementation.

From October 2021 onwards, Eurosolar4All developed energy communities targeted at vulnerable households in four locations across Southern Europe: the cities of Barcelona, Rome and Almada, in Spain, Italy and Portugal, respectively, and Coeur de Savoie, a peri-urban federation of municipalities (communauté de communes) near Chambéry in France. As a project, Eurosolar4All worked well within a larger European policy framework that promotes RECs as a potential solution to advance energy efficiency and combat energy poverty. Its framing was inspired by NYSERDA's successful pilot, and the US-based authority engaged with the European project team to share their experience and give initial inputs towards framing the pilots. This led to the first phase of policy mobility, shaping the four pilot projects in these European contexts by early 2022.

Thereafter, project partners in the four locations – a mix of local and regional public authorities, national and transnational civil society organizations, and research and training institutions – set a series of actions in motion. These included framing the scheme in specific ways consistent with national legislation and locally available funding streams to build community-scale solar plants; identifying – according to particular logics – low-income households as intended beneficiaries of the solar production based on virtual net metering; and promoting the scheme. By late 2023, all four pilot schemes had received formal approvals and had experienced a range of success in attracting relevant member households. They had also encountered and navigated numerous hurdles and delays along the way. Institutional bureaucracy is a key constraining element of policy mobility (Joosse and van Buuren 2023), but it also presents an invaluable window into understanding how an idea-from-elsewhere is reshaped to become a constituent part of a local assemblage that it reconfigures in the process.

We mobilize this dual policy mobility lens based on our experience conducting the social impact assessment for the Eurosolar4All project, on the documents and reports produced by the project team since its launch, and on a variety of discussions and interactions with project members. We were also involved in drafting the grant application and participated in various consortium meetings with the project team in all four locations and in most monthly online meetings over the course of three years. We undertook visits to the sites of the four pilot projects to understand the intended setups, associated challenges and the different sociocultural contexts.

Additionally, we conducted 18 interviews with prospective beneficiaries in Coeur de Savoie (January 2023, four interviews), Almada (June 2023, six interviews), Barcelona (October 2023, four interviews) and Rome (February 2024, four interviews). None of the prospective beneficiaries had yet received any benefits from the scheme, but they had all participated in training, workshops and/or events aimed to raise awareness and encourage new members to join the pilot projects. Moreover, we completed the first wave (of 2) of a structured survey towards social impact assessment, with a total of 149 respondents (12 in Coeur de Savoie, 61 in Almada, 52 in Rome and 24 in Barcelona). While we do not delve into the interview and survey results here, they give us a baseline understanding of the households targeted and the implementation process. These data collection tools focussed on capturing socioeconomic characteristics of participant households, their energy access practices and expenditure, experiences of energy poverty and their expectations of Eurosolar4All. We were able to establish on this basis that the selection was overall successful in targeting households in energy poverty (based on high proportions of respondents with high-energy expenditure, energy services-related discomfort in their own homes, and a range of practices to reduce home energy expenditure) and that these households had very limited knowledge of energy support schemes and energy efficiency.

Next, we analyse the dual policy mobility, as the project transformed from Solar for All into Eurosolar4All, and the latter's operationalization in these four pilot contexts. We are especially interested in how contextual aspects (understood as geographical specificities, actors, and regulations and policies) impact the governance of these carbon emission reduction projects. What can we learn from how European, national, regional and local dynamics impacted the scheme design and emergent implementation efforts in the different settings? Our hope is to offer insight into how a sociocultural understanding of carbon emission reduction schemes can foster inclusion and empowerment in future transitions to low-carbon energy systems.

First mobility: from Solar for All to Eurosolar4All

In terms of regulations and policies, the New York State Climate Leadership and Community Protection Act, signed in July 2019, includes ambitious targets for renewable energy deployment, including that of reaching 70% of electricity production from renewable energy sources by 2030 and 6 gigawatts (GW) of distributed solar power by 2025. It also advocates for "enhanced incentive payments for solar and community distributed generation projects, focusing in particular but not limited to those serving disadvantaged communities" (New York State 2019, §66-p) – a similar objective to the one promoted by the European RED II directive. As a representative of NYSERDA underlined,[3] community solar in both New York State and Europe is considered more flexible and scalable than rooftop solar and provides a suitable solution to ensure participation and engagement from a diversity of individuals and households, including those living in collective buildings and high-density areas such as New York City.

Though initiated and launched before the Climate Act was passed, the Solar for All programme reflects its spirit. It creates community solar for vulnerable households, and it was conceived in response to some of the barriers that low-income households might face in joining an energy community. Largely owned and managed by private actors, existing solar communities often require members to pay a monthly subscription and undergo a credit check. In contrast, the NYSERDA solar energy communities, though built and operated by private companies, can be joined with no upfront cost and provide monthly savings to their members, from a few dollars in the winter months and up to 15–25 dollars in the summer months. To sign up participants, who may live anywhere in New York State, NYSERDA partnered with local organizations, including civil society organizations and housing providers. The initial objective of NYSERDA was to set up 20 community solar projects, which would cater to 10,000 low-income households.

However, NYSERDA encountered difficulties signing up participants, as households often lacked familiarity with the concept of community solar, did not want to undergo the arduous application process, or associated the project with predatory energy retailers (Cleto and Gomes, 2022).[4] After three years, they had signed up 4,000 participants out of the 10,000 initially planned and had set up nine solar communities. No community engagement activities were organized. The initial Solar for All programme recognized the difficulties faced by energy-poor households and provided them with lower energy tariffs and reduced barriers to join an energy community. It did not, however, engage directly with households, nor did it involve them in decision-making processes or allow them to actively participate in the energy community. By 2021, NYSERDA was looking to scale up the Solar for All programme and hoped to reach more than 160,000 low-income households. In response to the challenges they faced in signing up participants, they planned on instituting automatic enrolment. Instead of engaging with potential beneficiaries' understandings of the energy sector and its related sociocultural complexities, they, therefore, opted for a "technical" fix, which would also lead them to move away from the concept of energy communities as understood in the European Union (EU), which includes "open and voluntary participation" (Hanke et al., 2021). For this new phase of the Solar for All programme, and responding to complaints from first-phase participants discussed with Eurosolar4All during a seminar early on in the project, NYSERDA considered integrating Solar for All with other benefit programmes and levelling monthly savings to spread the benefits throughout the year, as this was one of the main complaints of first-phase participants.

At its launch in 2021, the Eurosolar4All project emulated many aspects of the Solar for All programme. Based on a top-down approach to planning and design, it aimed to create RECs for vulnerable households. Grounded in recently approved supportive European regulatory frameworks, it aimed to lower barriers to energy communities and to provide material benefits (lower energy tariffs) to energy-poor households in diverse urban and peri-urban contexts. The Eurosolar4All project also featured several distinct specificities: first, it was funded by a Coordination and Support Action grant through the Horizon 2020 framework programme, which informed its functioning (regular reporting obligations, key performance indicators and suchlike) and limited its timeframe (three years up to September 2024) and its ambitions to work with the available budget and human resources.

Thus, Eurosolar4All aimed at providing support to 400 households. Pilot cases were spread across four national legislations and local jurisdictions, which meant that, contrary to the Solar for All programme, which was implemented by a singular entity within a singular regulatory framework, Eurosolar4All was run by a consortium of diverse actors. Therefore, the design of community-scale solar interventions had to be adapted to different regulatory systems and governance structures. Eurosolar4All focussed on the local scale and involved municipal actors, whereas Solar for All was planned and designed at the state level. Eurosolar4All was set up as an experimental project: one of its main objectives was to provide a blueprint for replicability in diverse European contexts. Scaling up and expansion were therefore planned based on the results of the pilot schemes and by building up a community of practice where other municipalities and actors could learn from the experiences.

Finally, the Eurosolar4All project included energy efficiency advice and community participation and involvement. The objective of the project was to familiarize beneficiaries with the energy transition and to develop a feeling of engagement and ownership in their REC. Thus, whereas Solar for All focussed on material benefits and sought to scale up carbon emission reductions, the Eurosolar4All project adopted a more holistic approach to energy poverty alleviation, which included engagement with and empowerment of energy-poor households. This, however, meant scaling up could be a slow process, as the limited number of beneficiaries targeted by Eurosolar4All meant its replication elsewhere was highly reliant on the success of these local pilot projects.

Table 1.1 summarizes the main similarities and differences between the Solar for All and Eurosolar4All programmes. Analysed through a policy mobility lens, with specific attention to place and context in policy definition, this table brings to the fore the role played by sociocultural dynamics in the translation of Solar for All into Eurosolar4All. On the one hand, the European funding that enabled the implementation of Eurosolar4All also constrained and informed it, impacting the scope and timeframe of the project and prompting greater attention to local specificities to ensure replicability across the

Table 1.1 Comparison between Solar for All and Eurosolar4All

Similarities between Solar for All and Eurosolar4All	Differences between Solar for All and Eurosolar4All
RECs for vulnerable consumers, based on supportive regulatory frameworks.	Scale of implementation (state/region versus local/urban level).
Targeted at households.	Scope (timeframe and number of beneficiaries).
Top-down approach to planning and design.	Unity/diversity of regulatory and governance contexts.
Financial benefits for participants.	Overall philosophy of project (scaling up versus experimentation for replicability).
Diverse urban and peri-urban contexts.	Modes of engagement (financial versus community participation and involvement).
	Sociology of actors involved in policy definition (state versus local level and public actor versus diversity of actors, including civil society organizations and research and training institutions).

EU. On the other hand, the greater emphasis on building communities through sustained interactions with beneficiaries, as well as on beneficiary empowerment and training, is rooted in the visions, experiences and priorities of the actors involved in drafting the Eurosolar4All project, including locally engaged civil society organizations as well as research and training institutions. We elaborate upon this in our analysis and discussion below.

Second mobility: adapting Eurosolar4All to different national and local contexts

Next, for each Eurosolar4All pilot locality, we present the overall context (geographical specificities, actors, regulations and policies) (Table 1.2), as well as the variations on the overall design for the first phase of the project (Table 1.3). The different actors involved in the design and implementation of Eurosolar4All were deeply aware of the differences between the state of New York and their own contexts. They were also aware of the differences across the four European sites. This influenced how they adapted the project for implementation in different contexts. Whereas Barcelona (Spain) and Almada (Portugal) both opted for collective self-consumption models for specific buildings, Rome and Coeur de Savoie designed their own models. Each pilot also adopted different strategies to recruit and engage beneficiaries, based on their understanding of local dynamics and their institutional and financial capabilities (Bonsfills and Rodríguez 2022).

Barcelona, Spain

Barcelona, the capital of the autonomous region of Catalonia, is the second-largest metropolitan area and one of the most populous cities in Spain, with

Table 1.2 Specificities of each context

	Geographical specificities	Actors	Regulations and policies
Barcelona	Regional capital with a very high population density.	Non-profit energy consultancy firm and Municipal Energy Agency. Extensive experience with projects funded by European agencies.	Basic definition of RECS. More detailed regulations regarding collective self-consumption.
Almada	Medium-sized municipality within the metropolitan area of Lisbon.	Municipal Energy Agency and Municipality. Extensive experience with projects funded by European agencies, but recent organizational restructuring.	Decree 162 of 2019: Legal Framework for Renewable Energy Communities and Collective Self-Consumption.
Rome	National capital with a very high population density.	Research and training institution and Municipality.	Law Decree n.119 8-11-2021.
CCCS	Peri-urban areas in a mountainous region with a low population density.	Research and training institution and Municipality.	Awaiting decree specifying modalities for the establishment of RECs.

Table 1.3 Departures from basic design

	Design for phase 1	Advantages/disadvantages
Barcelona	Collective self-consumption in a single social housing building coupled with an intensive engagement strategy.	Unity of place facilitates engagement and participation but limits the number of beneficiaries and administrative hurdles related to collective self-consumption.
Almada	Collective self-consumption in a single social housing building coupled with an intensive engagement strategy.	Unity of place facilitates engagement and participation but limits the number of beneficiaries and administrative hurdles related to collective self-consumption.
Rome	Selection of beneficiaries based on a city-wide database. Proximity to the solar PV plant as a criterion for selection. Increased number of beneficiaries due to local political impetus.	More geographically diverse pool of beneficiaries. Larger number of beneficiaries. Risk of low engagement and participation of households.
CCCS	Focus on energy renovation to improve energy efficiency. Integration of benefit programmes.	Offers a more comprehensive approach to energy transition but limits control over selection. Risk of low engagement and participation of households.

more than 1.6 million residents. A total of 11% of the city's population suffers from energy poverty (Tirado-Herrero 2022). Barcelona Municipality's 2030 Climate Plan also underlines wide discrepancies in levels of energy poverty depending on the neighbourhood. In Spain, the regulatory framework currently only provides a basic definition of RECs. In terms of collective self-consumption, regulations are more established. Among relevant aspects, we can underline that collective self-consumption must take place within a radius of 500 metres and it does not require the creation of a legal entity (Cleto and Gomes 2022). Locally, the Eurosolar4All pilot is managed by EcoServeis – a non-profit consultancy firm – and by the Agència Local d'Energia de Barcelona, the municipal energy agency. Both institutions have experience working together and have extensive experience with leading and contributing to European projects.

For the first phase of the project, partners in Barcelona decided to focus on recently built and energy-efficient public social housing. Solar PV panels were installed on the roofs of the buildings. A total of 12.5% of the electricity produced was to be used to supply common areas and infrastructures, while each resident would receive a 2.5% share of the electricity produced. The financial benefit would thus take the form of monthly savings on their electricity bill. This collective self-consumption model had some advantages and disadvantages. The proximity to the solar power plant facilitated the engagement and participation of beneficiaries. It also facilitated the involvement of Eurosolar4All partners, who were able to put up posters in the building, distribute flyers and organize informative sessions (Alves and Bonsfills 2022). Partners in Barcelona also favoured a "familiarity" approach to engage beneficiaries, including ensuring a single contact person, one-to-one meetings (either in person or online), personal phone calls and the possibility for beneficiaries to contact the Ecoserveis team through the application WhatsApp (Bonsfills and Rodríguez, 2022). This emphasis on interaction and familiarity was visible during one author's visit to the buildings, during which the accompanying contact person from Ecoserveis, who was also the single contact person for the building, spent time answering questions from residents, and later – during a discussion with the author – emphasized the importance of listening to residents' queries and building trust. The Barcelona partners also benefitted from the support of the Social Housing Agency and from the building managers. However, in the second building, tenants were only able to move in from March 2023 onwards. This limited the initial number of beneficiaries (35), though 41 tenants from the second building eventually joined the first phase (Bouvier 2023). This model also presented challenges to implementing collective self-consumption, including a complex administrative process and numerous interactions and discussions with the electricity distribution utility companies. This led to delays in the disbursement of the financial benefits and, therefore, to demotivation among certain households, as the benefits of the project were slow to emerge. This complexity illustrates the relevance of understanding sociocultural dynamics and how they can impact low-carbon transition efforts in practice.

Almada, Portugal

Almada is a municipality situated within the Lisbon metropolitan area, located across the Tagus River from Portugal's capital. With a population of around 170,000 people, the city is under urban pressure from Lisbon and hosts an important number of social housing buildings, which are managed by the municipality. Portugal has one of the highest rates of energy poverty in Europe. According to Eurostat (2020), 17% of the Portuguese population is not able to adequately heat their home. In terms of regulations and policies, Portugal first introduced a legal framework for Clean Energy Communities in 2019 (Scharnigg and Sareen 2023). Decree 162 of 2019 recognizes both RECs and collective self-consumption; the main difference between them being that collective self-consumption does not require the creation of a legal entity and allows for profit-making, which is not the case for the RECs. The regulatory framework is further clarified in Decree 15 of 2022, which details the concept of proximity between the place of energy generation and the place of consumption. Locally, the Eurosolar4All project is managed by Ageneal, the municipal energy agency, which underwent important organizational restructuring in 2022, and by the Camara Municipal de Almada, the municipality. Ageneal has extensive experience contributing to European projects.

As in Barcelona, the Almada pilot favoured a collective self-consumption model for the first phase of implementation and targeted social housing buildings where households would benefit from monthly savings on their electricity bill (Alves and Bonsfills 2022). Having the solar panels placed atop the social housing where beneficiaries lived was not only a practical arrangement from a legal and economic perspective (the roof was available and the location would qualify for eligibility within formal definitions of RECs), but the partners also saw this as an important step in terms of engaging households and helping to cultivate a sense of ownership. They took pains to engender such a sense but also encountered practical challenges in doing so, especially as there were no immediate benefits to be disbursed or pre-installed solar panels or subsidized solar energy tariff lines on the beneficiaries' electricity bills. During a project meeting one author attended in Almada in June 2023, the local project partners explained that the delays – and limited benefits per household even once the pilot was implemented – meant it was important to show the beneficiaries that something concrete was actually taking place. To facilitate engagement, local partners focussed on social housing residents living in or in the vicinity of a building in which they had already conducted an energy renovation project. In terms of recruitment and engagement, they also adopted a "trust-based" approach by ensuring that informative sessions and workshops would be held close to the beneficiaries' building to facilitate participation and by prioritizing in-person interaction with them (Bonsfills and Rodríguez 2022). The question of what trust-based means is a complex one, and understandings varied between project partners, as well as between them and beneficiary households. These understandings were also mediated by the practicalities of implementation:

project partners had limited time to conduct this work and residents had limited time to participate in workshops without the promise of immediate tangible benefits, which naturally constricted engagement. Partners from Almada faced similar issues to those in Barcelona in terms of implementation: while they had no problems with engagement and participation, they were nonetheless limited in the number of beneficiaries that could participate due to the envisaged solar capacity on offer, thus only ten households joined in the first phase. They encountered difficulties finding funds for the installation of the solar rooftop and delays in activating self-consumption. They planned to recruit 90 households in the second phase, and in July 2023, they received formal approval from the national authority to proceed with the pilot, with funding for the solar modules also included within the internal budget.

Rome, Italy

Rome, the national capital of Italy, is also one of the largest metropolitan areas of the country and one of its most populous cities – the municipal area has more than 2.8 million residents. According to Eurostat, in 2020, 8% of the Italian population was unable to keep their home adequately warm.[5] With regard to the regulatory framework, decree-laws passed in recent years have provided details on collective self-consumption and RECs. This includes Law Decree 119 8-11-2021, which transposes the RED II directive into Italian law (Cleto and Gomes 2022). Locally, Eurosolar4All is managed by the Department of Social Policies and Health, Roma Capitale, and by the Interdepartmental Research Centre for Territory, Building, Heritage and Environment (CITERA) at Sapienza University, which is involved in several European energy projects.

Contrary to the partners from Barcelona and Almada, the partners from Rome looked past the possibility of focussing on specific buildings and instead decided to identify households based on two criteria: whether they benefitted from the Bonus Energia 2019 (a targeted energy subsidy) and whether they lived in proximity to a municipal solar PV plant. Other indicators included social housing and low-quality buildings, as well as the presence of local actors (such as neighbourhood welfare associations) that could support participant engagement (Cumo et al. 2022). They thus used a city-wide database to select beneficiary households, customizing the approach based on locally situated factors. These factors included their organizational and data infrastructures and thus were somewhat removed from local realities, as they were based on data infrastructure categories rather than on potential participant households' lived realities. This desk-based approach nonetheless helped ensure systematic and potentially scalable identification of households in energy vulnerability, but it also proved to be a very time-consuming exercise due to bureaucratic practices and municipal protocols, for example, relating to collecting and matching beneficiary documentation. This time-intensive aspect led to challenges in terms of household engagement and participation. A proposed solution to this issue was to focus on solar plants built in schools, with the hope of then being able to rely on the school community to facilitate engagement. The Rome partners' engagement strategy also included WhatsApp groups and email lists, as well as collaborating with other local actors to foster interactions with beneficiaries (Bonsfills and Rodríguez 2022). However, in one of their project reports, Rome's partners underscored low attendance at their workshops and a general lack of motivation from beneficiaries as challenges to implementation (Bouvier 2023). Another particularity of the Rome pilot was a more ambitious target in terms of beneficiaries. As Cumo et al. (2022, 6) underline,

> the initial proposal foresaw two cycles for 50+50 beneficiaries, but several operational aspects, including management procedures, and the intention to have an impact proportioned to the size and resources of the city, forced an increase in the number of beneficiaries to 200 vulnerable households.

Thus, the Roman case makes evident the importance of local political impetus in enlarging the scope and design of the project.

Communauté de communes Coeur de Savoie, France

The Communauté de communes Coeur de Savoie (CCCS) is an administrative region that includes 41 municipalities and villages in a peri-urban area in the vicinity of Chambéry in the French Alps. It has a population of around 37,000. This reflects a much lower population density than the other pilot localities. Another specificity of CCCS is that housing is mostly dispersed, with 74% standalone houses, contrary to the predominance of collective buildings in the other Eurosolar4All pilots. According to the Observatoire National de la Précarité Énergétique, nearly 12% of French households experienced energy poverty in 2020, with wide disparities across the territory. With regard to the regulatory framework, the European definition of RECs was transposed into French law in early 2021. A decree specifying the characteristics and modalities for the establishment of RECs was announced in November 2021 but still had not been released two years later. In terms of collective self-consumption, the regulatory framework was more established, with details regarding tariffs, capacity and proximity between points of generation and consumption (Cleto and Gomes 2022). Locally, the Eurosolar4All project was managed by the Institut National de l'Énergie Solaire (INES), which conducts research on solar energy and provides training, and the Federation of Municipalities CCCS, which has around 200 employees and thereby a much smaller institutional structure than the other municipal agencies involved in Eurosolar4All. However, CCCS has a long history of developing solar projects at the local level, especially in the municipality of Montmélian, which installed solar rooftops and heating systems as early as the 1980s.[6]

The first phase of the French pilot was influenced by the Négawatt scenario,[7] which prioritizes energy sufficiency and efficiency along with renewable energy deployment. This scenario advocates for a more holistic approach to energy transition that would lead to suppressing energy waste, rethinking urban models, renovating buildings, increasing the energy performance of appliances and developing renewables. Inspired by this approach, the French partners decided to partly integrate the local Eurosolar4All pilot into subsidies for energy renovation. They recruited participating households through two main channels: first, social workers were asked to identify households in situations of energy vulnerability within the CCCS territory. Second, households in the CCCS territory who had started a house energy renovation project (such as wall insulation or upgrades to their boiler or heating system) and were eligible for the French state subsidy (based on taxable income) were offered an additional subsidy. In both cases, households benefitted from a one-time payment of several hundred Euros and from individual energy advice sessions organized by CCCS and/or a local association acting as a contractor for the municipality. However, this pilot faced several challenges, such as slow participant recruitment due to their highly dispersed nature, a lack of control over the selection of beneficiaries (as with the Almada case, these were selected based on criteria defined by the national government) and difficulties in ensuring the participation and involvement of households in planned activities. Though workshops were planned, as in the Roman case, the absence of a co-located community complicated household engagement, as this required time, transport and coordination on the part of users and stretched the limited human resources of the project. Our empirical material offers insight into the organizational challenges and dynamics at a stage when users were being enrolled rather than engaged as active participants and makes a crucial point in relation to how low-carbon transitions can be implemented in socially and culturally inclusive ways. As intermediaries and facilitators, project partners co-constitute the sociocultural nature of carbon in the very process of bringing it into being with households in these contexts.

As summarized in Table 1.3, the design of the first phase of Eurosolar4All varied significantly across the four pilot projects. In the next section, we turn to a concluding discussion of the sociocultural dynamics that informed these variations and summarize what Solar for All and EuroSolar4All can tell us about the different paths to a more inclusive and just energy transition.

Discussion and conclusion

The case of Eurosolar4All provides insights into the role that sociocultural dynamics play in the design and implementation of RECs as a socially orientated modality of carbon emission reduction for energy-poor households. We highlight three characteristics in the following paragraphs.

First, we highlight the role of human actors and their visions, experiences and priorities in socioculturally shaping carbon emission reduction. In the Iberian cases, actors with extensive experience implementing pilot projects through a European grant chose the path of least resistance (collective self-consumption, which could evolve into a REC, for which the Almada case did eventually receive formal approval in July 2023 after national level bureaucratic delays). They combined this with a time- and work-intensive engagement strategy, based on cultivating familiarity and trust. This strategy was informed by their understanding of the beneficiaries' mobility constraints and socioeconomic difficulties. These choices helped them to ensure that key performance indicators and objectives were met during the span of the project. Issues of internal organization (such as major organizational restructuring in Almada or a large and bureaucratic social department with a heavy workload in Rome) did exert their influence on the implementation timeline, nonetheless. Delays resulting from administrative and regulatory hurdles similarly impacted the motivation of beneficiaries (Bouvier 2023). It is also worth noting the role of existing institutional relationships and proximity, such as the mutually supportive interaction and cooperation of Eurosolar4All partners with social housing agencies and social work departments at the local level. By contrast, interactions with electricity distribution utility companies at the regional and national scale proved considerably more challenging.

Second, in terms of context-specific opportunities and constraints, we note the importance of political and regulatory frameworks at different scales in how carbon emission reduction is approached in policy and implementation processes, as these condition and structure the very space of possibility for citizen engagement with carbon emission reduction schemes. This arguably necessitates the involvement of intermediaries to enable implementation and involvement, and thereby it is important for scholarship on the sociocultural composition of carbon reduction schemes to bring these institutional actors into its purview. In most European countries, the absence of a clear national legal definition and regulatory framework for RECs tends to prompt a preference for collective self-consumption rather than the development of actual RECs as defined according to the RED II directive. In CCCS, the influence of the Négawatt scenario at the local level impacted the pilot project design, with a specific focus placed on energy efficiency.

Third and last, we underscore the role of geographical specificities, and especially of population density and housing characteristics, in determining the different paths to carbon emission reduction.[8] These pathways evolved during implementation. Thus, carbon emission reduction could not be simply specified as the amount of solar energy generated and consumed by beneficiary households in the pilot RECs, displacing grid electricity that has a higher level of associated carbon intensity. In the French case, the predominance of scattered individual housing and homeowners complicated user engagement but facilitated the development of a model based on energy renovation. In Rome, local political impetus relating to the size of the city led to a substantial increase from the initially targeted level of diffusion.

These insights into the sociocultural constitution of carbon emission reduction towards serving a synergistic social goal of pro-poor solar rollout can help understand the differences in how the Solar for All programme and

Eurosolar4All project engaged with energy poverty alleviation. This in turn can help us learn how to foster inclusion and empowerment through such interventions and what challenges to plan for. First, we emphasize the implications of the differences between the Solar for All and Eurosolar4All models. The latter focussed more extensively on the engagement and involvement of prospective beneficiaries yet faced significant and varied hurdles in implementing various pilot project approaches, despite trying to make them fit-for-purpose within each context. This experience was symptomatic of the tension between the need to rapidly scale up renewable energy adoption and the aspiration to engender a sense of community and generate knowledge for project replicability elsewhere. Attention to this sociocultural tension raises the question of whether to prioritize a large-scale redistribution of the material benefits of a rapid energy transition, which is not likely considering capitalist modes of financing and implementation, or to invest time in small-scale interventions with models of user engagement and empowerment that can be besought with bureaucratic and technocratic challenges. This chapter shows that there are no easy answers, and yet that this itself is a valuable lesson learnt from unpacking the sociocultural contours of carbon reduction schemes.

Second, the four Eurosolar4All pilot projects constitute different paths. The Iberian model promotes active engagement at the building scale, which can eventually be scaled up by collaborating with social housing agencies and utility companies at the municipal level or lead to the establishment of actual RECs. The French case indicates a path towards increasing synergy between energy efficiency measures, renewable energy development and energy poverty alleviation by integrating a panoply of benefit programmes – an approach that may be more suitable for smaller municipalities and peri-urban areas. The Roman case indicates yet another path for large European cities, namely to direct profits made from municipal solar plants to vulnerable households in a systematic manner. In this sense, the Rome example is closest to the Solar for All programme.

Finally, although the Solar for All and Eurosolar4All projects offer many different and potentially rich paths to synergize energy poverty alleviation and renewable energy deployment at the local and regional scale, it is apparent from our study that in all the case studies, the design and implementation processes were highly top-down and far from user-led. The inclusion of vulnerable households in energy communities and transitions was mostly envisaged in the form of financial benefits and awareness-raising workshops, in line with a long-running energy sector trend (Komendantova et al., 2018). A remaining challenge, therefore, is how to promote RECs that are actually designed and controlled by vulnerable households – a core aspect of the European Green Deal that seeks to place citizens at the centre of societal energy transitions and foresees a significant role for energy communities within such a vision. Despite the best of intents and intensive applications, the dual policy mobility from Solar for All to the four Eurosolar4All pilot projects suggests that the centralized, consumer-oriented – rather than user-centric – constitution of carbon embedded in everyday energy infrastructures remains largely impervious to attempts to advance towards a just transition. A better understanding of the sociocultural dynamics entailed in such a shift has both conceptual and practical relevance.

Acknowledgements

The authors acknowledge the Horizon 2020 funded Sun4All project (grant 101032239), the Research Council of Norway funded ASSET project (grant 314022) and SOLNOR project (344206), and the JPI Urban Europe (DUT Call) and Research Council of Norway-funded ENERGY4ALL project (grant 349994), for funding time that made this research possible. We are also grateful to the edited volume editors for organizing a workshop in Copenhagen in April 2023 which helped to ensure thematic cohesion across chapters, and to the Sun4All project participants for the time and insights that they shared.

Notes

1 The European Commission specifies that "Energy poverty occurs when a household must reduce its energy consumption to a degree that negatively impacts the inhabitants' health and wellbeing. It is mainly driven by 3 underlying root causes: a high proportion of household expenditure spent on energy; low income; low energy performance of buildings and appliances." See https://energy.ec.europa.eu/topics/markets-and-consumers/energy-consumer-rights/energy-poverty_en, last accessed on May 27, 2024.
2 An overview of the Eurosolar4All project is available here: https://cordis.europa.eu/project/id/101032239 (last accessed on May 27, 2024).
3 All the information on Solar for All presented in this section is based on seminars and presentations made by a representative of NYSERDA and on Eurosolar4All Deliverable D2.1, a report entitled: "Blueprint model for the Sun4All programme" (Cleto and Gomes, 2022).
4 For more details, see the project website: https://www.nyserda.ny.gov/All-Programs/NY-Sun (last accessed on May 27, 2024).
5 For details, see national statistics on the Eurostat website: https://ec.europa.eu/eurostat (last accessed on May 27, 2024).
6 See https://france3-regions.francetvinfo.fr/auvergne-rhone-alpes/savoie/depuis-30-ans-montmelian-rayonne-l-energie-solaire-870763.html (last accessed on May 27, 2024).
7 The Négawatt scenario is produced by a French non-governmental think tank, which brings together experts from different fields related to energy. The scenario discusses the future of the French energy sector and possible pathways to sustainability. It advocates for a greater focus on energy sufficiency and efficiency.
8 Here, carbon emission reduction is to be calculated once implementation is completed, based on the solar generation and consumption across beneficiary households. However, in the pilot implementation phase that our empirical material draws upon, this stage had not yet been reached. Hence, to be true to our material, we capture the interim dynamics, which we also see as being generative of relevant insights.

References

Aiken, G. Taylor. 2017. "The Politics of Community: Togetherness, Transition and Post-politics." *Environment and planning A: Economy and Space* 49, no. 10 (August): 2383–2401. https://doi.org/10.1177/0308518X17724443

Alves, Catarina, and Mariona Bonsfills. 2022. *Implementation Plan of Sun4All Programme*. Almada: Eurosolar4All. Accessed on November 21, 2023, URL: https://sunforall.eu/resources

Bielig, Mona, Celina Kacperski, Florian Kutzner, and Sonja Klingert. 2022. "Evidence Behind the Narrative: Critically Reviewing the Social Impact of Energy Communities in Europe." *Energy Research & Social Science* 94, (December): 102859. https://doi.org/10.1016/j.erss.2022.102859

Birch, Kean. 2017. "Techno-economic Assumptions." *Science as Culture* 26, no. 4 (October): 433–444. https://doi.org/10.1080/09505431.2017.1377389

Bonsfills, Mariona, and Raquel Rodríguez. 2022. *Local Work Plans of Community Work*. Barcelona: Euroslar4All. Accessed on January 8, 2023. URL: https://sunforall.eu/resources

Bouvier, Xavier. 2023. *Monitoring Report on Implementation*. Cœur de Savoie: Eurosolar4All. Accessed on November 23, 2023. URL: https://sunforall.eu/resources

Bouzarovski, Stefan, and Saska Petrova. 2015. "A Global Perspective on Domestic Energy Deprivation: Overcoming the Energy Poverty–Fuel Poverty Binary." *Energy Research & Social Science* 10, (November): 31–40. https://doi.org/10.1016/j.erss.2015.06.007

Capstick, Stuart, Irene Lorenzoni, Adam Corner, and Lorraine Whitmarsh. 2014. "Prospects for Radical Emissions Reduction through Behavior and Lifestyle Change." *Carbon Management* 5, no. 4 (April): 429–445. https://doi.org/10.1080/17583004.2015.1020011

Chilvers, Jason, and Noel Longhurst. 2016. "Participation in Transition(s): Reconceiving Public Engagements in Energy Transitions as Co-Produced, Emergent and Diverse." *Journal of Environmental Policy & Planning* 18, no. 5 (January): 585–607. https://doi.org/10.1080/1523908X.2015.1110483

Clarke, Nick. 2012. "Urban Policy Mobility, Anti-Politics, and Histories of the Transnational Municipal Movement." *Progress in Human Geography* 36, no. 1 (July): 25–43. https://doi.org/10.1177/0309132511407952

Cleto, João, and Pedro Gomes. 2022. *Blueprint Model for the Sun4All Programme*. Almada: Eurosolar4All. Accessed on November 21, 2023. URL: https://sunforall.eu/resources

Cochrane, Allan, and Kevin Ward. 2012. "Researching the Geographies of Policy Mobility: Confronting the Methodological Challenges." *Environment and Planning A* 44, no. 1 (January): 5–12. https://doi.org/10.1068/a44176

Cumo, F., P. Maurelli, E. Pennacchia, and F. Rosa. 2022. "Urban Renewable Energy Communities and Energy Poverty: A Proactive Approach to Energy Transition with Sun4All Project." *IOP Conference Series: Earth and Environmental Science* 1073, no. 1, 012011. https://doi.org/10.1088/1755-1315/1073/1/012011

European Union. 2018. *Revised Renewable Energy Directive or RED II (EU/2018/2001)*. Brussels: EU.

European Union. 2019. *Directive on common rules for the Internal Electricity Market (EU/2019/944)*. Brussels: EU.

Hanke, Florian, Rachel Guyet, and Marielle Feenstra. 2021. "Do Renewable Energy Communities Deliver Energy Justice? Exploring Insights from 71 European Cases." *Energy Research & Social Science* 80, (October): 102244. https://doi.org/10.1016/j.erss.2021.102244

Hedberg, Trevor. 2018. "Climate Change, Moral Integrity, and Obligations to Reduce Individual Greenhouse Gas Emissions." *Ethics, Policy & Environment* 21, no. 1 (March): 64–80. https://doi.org/10.1080/21550085.2018.1448039

Hoicka, Christina E., Jens Lowitzsch, Marie C. Brisbois, Ankit Kumar, and Luis R. Camargo. 2021. "Implementing a Just Renewable Energy Transition: Policy Advice for Transposing the New European Rules for Renewable Energy Communities." *Energy Policy* 156, (September): 112435. https://doi.org/10.1016/j.enpol.2021.112435

Jiglau, George, Stefan Bouzarovski, Ute Dubois, Marielle Feenstra, João P. Gouveia, Katrin Grossmann, Rachel Guyet, et al. 2023. "Looking Back to Look Forward: Reflections from Networked Research on Energy Poverty." *iScience* 26, no. 3 (March): 106083. https://doi.org/10.1016/j.isci.2023.106083

Joosse, Hans, and Arwin van Buuren. 2023. "The Marginalization of Policy Integration: Dynamics of Integrated Policymaking in the Periphery of Bureaucracy." *Public Policy and Administration* 39, no. 3 (June): 09520767231175917. https://doi.org/10.1177/09520767231175917

Komendantova, Nadejda, Monika Riegler, and Sonata Neumueller. 2018. "Of Transitions and Models: Community Engagement, Democracy, and Empowerment in the Austrian Energy Transition." *Energy Research & Social Science* 39, (May): 141–151. https://doi.org/10.1016/j.erss.2017.10.031

MacArthur, Julie L. 2016. "Challenging Public Engagement: Participation, Deliberation and Power in Renewable Energy Policy." *Journal of Environmental Studies and Sciences* 6, (September): 631–640. https://doi.org/10.1007/s13412-015-0328-7

Nader, Laura. 1972. "Up the Anthropologist: Perspectives Gained from Studying Up". In *Reinventing Anthropology*, edited by D. Hymes, 284–311. New York: Pantheon.

New York State. 2019. *N.Y. Pub. Serv. Law § 66-P. Establishment of a renewable energy program*. Accessed on November 21, 2023. URL: https://casetext.com/statute/consolidated-laws-of-new-york/chapter-public-service/article-4-provisions-relating-to-gas-and-electric-corporations-regulation-of-price-of-gas-and-electricity/section-66-p-establishment-of-a-renewable-energy-program

Nielsen, Kristian S., Kimberly A. Nicholas, Felix Creutzig, Thomas Dietz, and Paul C. Stern. 2021. "The Role of High-Socioeconomic-Status People in Locking in or Rapidly Reducing Energy-Driven Greenhouse Gas Emissions." *Nature Energy* 6, no. 11 (September): 1011–1016. https://doi.org/10.1038/s41560-021-00900-y

Prince, Russel. 2017. "Local or Global Policy? Thinking About Policy Mobility with Assemblage and Topology." *Area* 49, no. 3 (December): 335–341. https://doi.org/10.1111/area.12319

Roberts, Mike B., Arijit Sharma, and Iain MacGill. 2022. "Efficient, Effective and Fair Allocation of Costs and Benefits in residential Energy Communities Deploying Shared Photovoltaics." *Applied Energy* 305, (January): 117935. https://doi.org/10.1016/j.apenergy.2021.117935

Sardo, Michael C. 2023. "Responsibility for Climate Justice: Political not Moral." *European Journal of Political Theory* 22, no. 1 (January): 26–50. https://doi.org/10.1177/1474885120955148

Sareen, Siddharth, Alevgul H. Sorman, Ryan Stock, Katherine Mahoney, and Bérénice Girard. 2023. "Solidaric Solarities: Governance Principles for Transforming Solar Power Relations." *Progress in Environmental Geography* 2, no. 3 (July): 143–165. https://doi.org/10.1177/27539687231190656

Scharnigg, Renée, and Siddharth Sareen. 2023. "Accountability Implications for Intermediaries in Upscaling: Energy Community Rollouts in Portugal." *Technological Forecasting and Social Change* 197, (December): 122911. https://doi.org/10.1016/j.techfore.2023.122911

Tirado-Herrero, Sergio. 2022. "Measuring Energy Poverty at the Urban Scale: A Barcelona Case Study." In *Energy Poverty Alleviation*, edited by Carlos Rubio-Bellido and Jaime Solis-Guzman. Cham: Springer. https://doi.org/10.1007/978-3-030-91084-6_13

2 Carbon footprint calculators and behaviour change

Quentin Gausset

Introduction

Since the end of the 19th century, the development and welfare of industrial countries have depended on the consumption of large quantities of fossil fuel. Industrial activities have released large amounts of greenhouse gases (GHGs) into the atmosphere. It is well known how this has triggered a climate crisis that threatens biodiversity and human livelihoods globally.

To tackle this global problem, international agreements such as the Paris Agreement, adopted in 2015, or the European Green Deal, approved in 2020, ask signatory states to rapidly reduce their carbon emissions and to reach carbon neutrality by 2050 to keep the rising global temperature below two degrees Celsius. Each state develops its own strategy to reach these goals, which tends to focus on replacing fossil fuels with renewable sources of energy and on encouraging people to consume more sustainably through a mix of new legislation and financial incentives (taxes and subsidies). Within states, an increasing number of cities, municipalities, enterprises and citizens also take steps to reduce their carbon emissions. Many of these entities consider that calculating carbon emissions is a necessary step to help achieve this goal (e.g. Knox 2020; Lippert 2015). One finds a plethora of carbon footprint calculators (CFCs) run by consulting companies to help, for example, businesses, offices or municipalities to measure their current emissions, set targets for reduction and monitor progress. For households or individuals, a number of apps or online CFCs try to make carbon footprint calculations accessible to everyone – and these are the focus of this chapter.

CFCs play a central role in addressing the climate crisis by helping a variety of actors to assess and calculate their current carbon emissions, establish a baseline, set reduction targets and monitor their progress to reach them. Thereby they establish a social understanding of emissions as footprints, and for all practical purposes, our actions come to depend on these calculations (e.g. Lippert 2015). CFCs thus translate the complexities of climate change and climate change mitigation action into measurable phenomena. Such translations are part of the cultural complexity of carbon.

DOI: 10.4324/9781003478669-3

CFCs are used on a wide variety of scales, ranging from the macro at the international level to the micro-level individual actions. CFCs are imagined as tools to help reduce carbon emissions by establishing climate policies with precise targets, such as developing a carbon tax, establishing a system of personal carbon allowances,[1] determining the purchase of carbon offsets[2] or simply informing citizens about their consumption in the hope that such information will trigger behaviour change. This chapter focusses on CFCs available online or in apps that aim to raise awareness and promote alternative behaviour. It thereby contributes to a growing number of studies that focus on how CFCs are employed to further the green transition (e.g. Jones and Kammen 2011; Pandey et al. 2011; Edstrand 2016; Salo et al. 2016, 2019; West et al. 2016; Collins et al. 2018, 2020; Burgui-Burgui and Chuvieco 2020; Kok and Barendregt 2021; Enlund et al. 2023).

While these studies discuss the functions of CFCs, they provide no clear evidence of the actual impact of CFCs on changing behaviour and reducing the carbon footprints of respondents or users. Several CFCs sell carbon offsets (Yayzy, Greenly, Klima or Ellie), but offsetting one's carbon footprint is not the same as changing one's behaviour to reduce it. Some CFCs keep track of the deeds and challenges supposedly accomplished by respondents. For example, Earth Hero claims that 117,000 participants have accomplished 670,000 separate actions, thereby saving more than 10 million tons of CO_2. But these deeds are self-declared and unverifiable, as is much of the data entered into CFCs in the first place. Measuring the impact of CFCs on environmental behaviour would require calculating the carbon footprint of the same respondents repeatedly over time and isolating the impact of the CFC from other factors. To my knowledge, this kind of study has never been done. There is no clear connection between the use of CFCs and the practices they aim to change (Salo et al. 2019). Simply communicating carbon footprints is not sufficient to trigger behaviour change (Bjørn-Hansen et al. 2022). There is little documentation of the effects of CFCs on behaviour change, and only a small group of people – those motivated to change behaviour – is interested in calculating their carbon footprint (Dreijerink and Paradies 2020; Gram-Hanssen and Christensen 2012). Therefore, the behavioural impact of CFCs should remain hypothetical until further research is able to confirm or disprove it.

Despite the lack of clear evidence on the impact of CFCs on behaviour change, CFC designers take for granted that CFCs do make a difference, and they develop CFC apps with that explicit goal. The present chapter first presents the challenges of calculating carbon footprints and then discusses some of the psychological and economic theories that are implicitly used to design CFCs that focus on individual behaviour. Finally, it argues that collective action – that is, joining or organizing efforts to reduce climate impact with others – for example, through shared living – can trigger much deeper behaviour changes than individual action, and that CFCs could therefore affect wider environmental behaviour change if they were also designed to measure the average consumption of "green communities" (such as eco-villages, green neighbourhoods or organic food cooperatives)[3] and not just that of individuals.

Methods

The present chapter is based on two types of material. First, inspired by auto-ethnography, I personally tested more than 30 CFCs by accessing and exploring their options. Homepages and apps are not always maintained, and some are no longer accessible, such as the 1tonmindre.dk/ homepage developed for the COP15 summit in Copenhagen or MapMyClimate.dk. Some of the CFCs can be found online, such as Carbon Independent (carbonindependent.org/), WWF Footprint Calculator (footprint.wwf.org.uk/), 1.5 Degree Lifestyles (carbonfootprint.hi.is/), The Nature Conservancy (https://www.nature.org/en-us/get-involved/how-to-help/carbon-footprint-calculator/), Loop (https://www.loophome.app/net-zero/), MyClimate (www.Myclimate.org), Carbon Footprint (www.carbonfootprint.com/calculator.aspx), UK Energy Watch (https://ukenergywatch.co.uk/carbon-footprint-calculator/) and Climate Stewards (www.climatestewards.org). Other calculators are accessible through apps such as Carbon Neutral and CO_2 Meter, Carbon Tracker, Pathways, Greenly, CarbonVisualizer, Footprint, Earth Hero, PersonalCarbonFootprint (Future Planet), Ellie, Klima, Myfootprint, Pawprint, MyFootprint/SEB, FlightCO2Calculator, Meatless, CO2 Meter, Earth Rewards, Yayzy, TribalData or the Coop supermarket app.

Second, I coordinated the development of a CFC as part of the Collective Movements and Pathways to Sustainable Societies (COMPASS) research project, financed by the Velux Foundation between 2017 and 2021, to evaluate the impact of collective action on environmental behaviour (measured in carbon footprints) in green communities. To test the hypothesis that collective and communally based action could produce greater behaviour changes than individual action, we developed a quantitative survey that compared the carbon footprint of 1,018 representative Danes with 258 members of green communities. This method rested on two assumptions. First, we assumed that environmental behaviour could be measured through carbon footprints linked to daily consumption. Second, we assumed that membership in a green community could be taken as a proxy of collective action. The details of our hypothesis and the survey are documented in other publications (see Toft et al. 2021; Gausset and Jensen 2024). My purpose here is to discuss these results in relation to the assumptions about change implied in the development of CFCs more generally.

Inspired by the carbonindependent.org CFC, the COMPASS project developed a set of questions to evaluate the carbon footprint of respondents that covered their energy consumption at home, their transport, their food habits and diverse forms of consumption. Energy consumption was measured on the basis of the actual consumption of energy for heating (modulated by the type of energy source) and the actual consumption of electricity.

We accounted for transport-related carbon footprints either by asking directly about the amount of fuel respondents bought for their cars and motorcycles or by asking about the distance or time they spent in cars or other means of CO_2-emitting transport. We multiplied these figures by an average CO_2 emission, either per litre of fuel or per kilometre or hour travelled (depending on the means of transport). Emissions from buses, trains or the metro are available on the webpages of the Danish public transport companies. For car-related carbon footprints, we modulated the carbon footprint by the declared car-sharing practices of respondents (e.g. someone travelling 100% of the time alone is responsible for 100% of their car emissions, but someone travelling 50% of the time alone and 50% of the time with four other passengers is only responsible for 60% of the total car emissions). In Denmark, high-speed ferries are common, and they emit roughly as much CO_2 per hour as plane transport. With an increasing number of electric cars available, careful attention was paid to make sure that car-related electric consumption was not already accounted for in the home energy consumption section discussed earlier.[4]

Food-related carbon footprints were estimated by asking respondents about how much of their consumed food was organic, meat-free, seasonal and produced in Denmark, or fresh and not-processed. Respondents were also asked to estimate how much of their food was self-produced (e.g. from growing vegetables or fruits at home or raising chickens) or was saved from being thrown out through practicing dumpster-diving or buying food items close to or past the best-before date. The carbon footprint of food items self-produced or saved from being thrown away was considered null in the survey. Last, we asked about how much peels and leftovers were composted, we asked about the amount of food waste and we applied the formula of carbonindependent.org to calculate food-related carbon footprints.

Finally, we estimated the carbon footprint for general consumption of a range of material items (e.g. clothes and shoes, furniture, restaurants and hotels, tobacco and bottled drinks, electronic devices, culture and entertainment, medicine and cosmetics) by asking respondents to estimate how much money they spent per year on each category and by associating these answers with an average carbon footprint per category and per Danish kroner (the Danish currency) derived from the website carbonfootprint.com.

Challenges in calculating carbon footprints

Engaging with the challenges in calculating carbon footprints requires us to enter a calculative universe: in this utopian universe, the calculative machinery and its assumptions are deemed unproblematic – though social science research has shown how the premises of such calculations are contested and their enactment in practice is not under control (e.g. MacKenzie 2009). This chapter proceeds within the logic of this calculative universe, in which there are two broad approaches to calculating carbon emissions. The first calculates emissions of a geographical entity by looking at the production, import and export of both fossil and renewable energy within and across specific geographical boundaries (Strobel et al. 2016). The second approach is consumption based. It consists of calculating the carbon footprint of a myriad of consumption items (food, clothes, electronics, transport and so on). The carbon footprint of a person, family or community is then calculated by adding the carbon footprint of all items consumed by that person or by all members of the family or community. This chapter focusses on the second approach, and in what follows I review a number of the calculative assumptions and challenges met in the design of CFCs, and then I discuss how the CFCs, as tools that assume individual change, can be analysed through the perspectives of a range of popular economic and psychological theories.

What level of detail?

Calculating carbon emissions requires measures and estimates that can be difficult to determine, and numerous social science publications have documented how challenging – and problematic – it can be to establish measures, counts and estimates of emissions and then compare the resulting carbon footprint with others (e.g. Lippert 2018). Calculating a detailed carbon footprint is a process that can take days, if not weeks, depending on the number of details informing the calculation. At one end of the spectrum, one finds CFCs, which are highly detailed and can distinguish the carbon footprint of a piece of hamburger, a chicken wing or a slice of ham. At this level of detail, respondents need to know the precise nature of their daily consumption. They need to keep a detailed track of what they eat in each meal; they need to know the precise quality of their clothes, including the place where they have been produced and the sustainability of the supply chain. The precision provided by these calculators requires, therefore, a high investment in time, and this may lead to high numbers of people abandoning the exercise before they can complete it. Therefore, all CFCs put a limit on the number of details they ask from respondents and make a trade-off between useability and complexity (West et al. 2016). The Klima app (one of the most detailed CFCs) states, for example, that "To calculate your carbon footprint from small appliances, we use average values – otherwise the questionnaire would be exhausting."

At the other end of the spectrum, one finds calculators that aggregate numbers based on wide-ranging averages. These calculators can ask for the total number of square metres occupied in housing and, on that basis, provide an average estimation of emissions related to heating and electricity consumption, or the CFC can ask how many kilometres someone has driven in a car and provide a carbon footprint based on the average consumption of an average car category. What is gained by these calculators in terms of user-friendliness is lost in terms of specific detail and context. In between these two extremes, one finds a variety of calculators that try to strike a balance. They choose a limited number of broad indicators that provide respondents with enough details to make the exercise meaningful, without requiring too much time to go through them. Some also begin with a very broad estimate of the respondent's carbon footprint based on a few generic questions and then, at a later stage, allow the respondent to refine their carbon footprint estimation with a set of more detailed questions or options (Earth Hero, Earth Rewards, Klima, Yayzy or MyFootprint/SEB).

What data, what scope to include and what calculation method?

The data and calculation methods used by CFCs differ widely (Pandey et al. 2011; Birnik 2013; Mulrow et al. 2019; Scrucca et al. 2021). One finds almost as many ways of calculating individual carbon footprints as one finds CFCs (Strobel et al. 2016). Some CFCs (Greenly, MyFootprint/SEB, Yayzy, Coop or Loop) calculate a carbon footprint based on transaction data registered automatically from purchases paid for with a credit card, with a supermarket member card or data from energy companies – an approach that is increasingly popular (Andersson 2020; Barendregt et al. 2020). But most ask respondents to answer questions about their consumption patterns – and all have that option available. Some CFCs are based in a specific country, such as the United Kingdom or the United States, while others ask respondents to state their country of residence and use relevant national data to calculate their carbon footprint (Klima, Earth Reward, Yayzy, TribalData, FuturePlanet or carbon footprint.com). Some CFCs focus on transport (FlightCO2Calculator), home energy (Loop) or diet (Meatless), but most cover a wide range of consumption aspects and also include energy use at home, clothes, electronics and so on.

The widely referenced GHG Protocol (www.ghgprotocol.org) distinguishes between three scopes of emissions, depending on their sources. Scope 1 emissions are those that a consumer emits directly (e.g. from burning fuel when driving). The easiest way to reduce Scope 1 emissions is to reduce consumption. Scope 2 emissions come indirectly from the production of items whose production emits CO_2. To reduce Scope 2 emissions, one can, for example, change energy provider, or choose to consume items that have a lower carbon footprint than others. Scope 3 emissions come from societal infrastructure or from a total value chain and are thus outside of consumer control. Most calculators include Scope 1 and Scope 2 emissions since these can be reduced by changing consumption patterns. Some CFCs, however, also include Scope 3 emissions that derive from societal infrastructure. For example, to evaluate air travel carbon footprints, carbonindependent.org takes the official consumption of a plane travelling x km, using y tons of fuel and with z seats, with an average seat occupancy of 80%. It multiplies the amount of fuel per person by a factor of 3.12 to calculate the amount of CO_2 emitted per person (90 kg per hour). Since emissions in the high atmosphere have a much greater greenhouse effect, the carbonindependent.org CFC multiplies the emissions by a factor of 2 (180 kg per hour). It then adds 70 kg to account for the extraction and transport of crude oil, inefficiencies in refineries, aircraft manufacture and maintenance, staff training, airport construction, maintenance, heating, lighting, etc.

The resulting calculation of 250 kg CO_2 per person per hour of flying thus includes some Scope 3 emissions. A few CFCs do the same for car transport (adding emissions from road construction and maintenance). Some also go further on to divide the carbon footprint by the (average) number of persons driving together. Thus, different CFCs will, in theory, give different carbon footprints per kilometre driven in the same car or per hour travelled in an identical plane.

Carbon footprint calculators as tools for individual change

In spite of the diversity that characterizes CFCs, they all present themselves as tools that aim to trigger behaviour change in the people who use them. To achieve this goal, they use a series of schemes which, in my analysis, are implicitly based on psychological and economic assumptions and theories of behaviour change. I next review different studies and theories that have been influential in designing the CFCs described in this chapter.

Some CFCs focus on informing respondents about the private benefits of being more environmentally minded. This approach might be said to rely broadly on *Rational Choice Theory*, which goes back to Adam Smith and considers that people are rational actors trying to maximize their private benefit and minimize costs. Benefits can be presented in financial terms, for example, in terms of savings connected to reducing consumption or in terms of the return on green investments (e.g. house retrofitting or installing solar panels). When green investments have a payback time of less than 30 years, they are seen as sound investments – the shorter the payback time, the better the investment. The information provided by these CFCs often relies on technical specifications and data from lifecycle analyses, but also on wide assumptions regarding average patterns of use and consumption. Yet they nevertheless give an idea of how much money can be saved by changing behaviour. For example, carbonfootprint.com – an American CFC – informs respondents that drying clothes by hanging them instead of using a drier can save 68 USD per year, turning the heating down by 1 degree can save 8% of the energy bill (55 USD in my case) and filling the kettle with the exact amount of water needed to be boiled can save 30 USD per year. Most environmentally sound options when consuming energy, transport, food or material items lead to economic savings, and insisting on this aspect is expected to lead people to adopt environmentally friendly behaviour to save money. The problem with this approach lies in the so-called "rebound effect." For example, people can use the money they save to consume something else (e.g. a trip to an exotic destination), which might cancel all their carbon savings.

But "rational" behaviour (i.e. trying to maximize benefits and minimize costs) does not necessarily need to be framed in financial terms. It can also include health, social or collective benefits. Like Rational Choice Theory, the *Theory of Planned Behaviour (TPB)* also assumes that people behave rationally and try to make the right choices, but in a wider sense than just financial benefit. This theory entails that behaviour is influenced by attitudes (the favourable or unfavourable evaluation of certain behaviour), subjective norms (beliefs about how others will perceive behaviour), and the perceived control that an individual has on the ability to perform certain behaviour and achieve goals (Ajzen 1991; Fishbein 1963, 1967, 1968; Fishbein and Ajzen 1972, 1974).

The Rational Choice Theory and the Theory of Planned Behaviour hold that providing people with comprehensive information and feedback makes them more aware of what they do and helps them change their attitudes by making well-informed choices to maximize different kinds of benefits. In that spirit, the Carbon Choices app states, for example, that

> This app is designed to help youth (and adults) to become more aware of the effects of their personal choices on greenhouse gas emissions and climate change. As consumers build an understanding and awareness of their part in climate change, we hope that they will attend to their choices more diligently.

Thus, the core implicit assumption behind CFCs that focus on individual behaviour is that by informing respondents and giving them feedback about the consequences of their actions, the app will encourage them to maximize the positive and minimize the negative consequences of their choices. This is primarily achieved through calculating the user's carbon footprint, which is supposed to raise awareness and trigger behaviour change, at least if the carbon footprint is limited to Scopes 1 and 2 emissions.[5] It can be noted here that informing about the saving potential of behaviour change in kg CO_2 rather than in monetary terms assumes that the primary motivation for behaviour change is also altruistic and aims at delivering collective benefits, instead of being driven by private benefits. Many CFCs also simply teach or inform respondents about the impact of their behaviour on the environment, biodiversity or the climate – sometimes with pages of facts, explanations and quizzes.

Some CFCs provide feedback on the respondent's carbon footprint in visual terms, showing, for example, an increase in the biodiversity of a given landscape when more pro-environmental options are chosen to trigger emotions and influence attitudes. This feedback can also relate the visual to the harmful consequences of unsustainable consumption. Mapmyclimate, for example, shows how much of Denmark or Copenhagen would eventually be flooded if all citizens on Earth adopted the same consumption pattern as the average respondent. Likewise, CarbonVisualizer uses the respondent's cell phone camera to superimpose fog on the image displayed to reflect the density of their carbon footprint.

Providing respondents with a list of concrete recommendations to reduce their CO_2 emissions (or to save money) can help them feel that they have control over their progress and that behaviour change is within reach. Recommendations can either be generic for all respondents, such as cycling more, eating less meat (the Meatless app makes this easier by providing vegetarian and vegan recipes), replacing old light bulbs with LED lamps and so on (Yayzy, TribalData and WWF MyFootprint), or they can be specific and tailormade based on the respondent's answers (Footprint and Future Planet). For example, a calculator might avoid advising respondents to cycle more if they already cycle a lot, or it might refrain from telling vegan respondents that reducing meat consumption can significantly reduce their carbon emissions.

One of the problems with simply providing recommendations is that their impact is not always self-evident, and it might need to be described with some level of detail. How much are CO_2 emissions really reduced by shortening a daily shower by 5 minutes? What difference does it really make to wash clothes at a low temperature? Some CFCs provide information on the difference that behaviour change can make by providing a calculation of the emissions users allegedly save when doing an environmental deed (Earth Rewards, Carbon Tracker or PawPrint). For example, the PawPrint CFC tells us that baking one's own bread can save 620g CO_2, while attending a meeting online instead of flying promises to save 463 kg CO_2 and taking a shorter shower saves 47g. Other CFCs display the respondent's carbon footprint in real time after each new answer (CarbonIndependent.org, Yayzy or Klima).

Given that CFCs use different data sources and calculation methods, the promise of transparency can be mobilized to help mitigate scepticism. Providing a detailed carbon footprint per specific answer category is a solution that allows respondents to experiment with CFCs and see for themselves how much they can reduce their carbon footprint by changing their diet from average meat and dairy to becoming lacto-vegetarian, for example (carbon independent.org). By doing this kind of exercise, people can identify areas where their behaviour change can have the biggest impact, which can help them prioritize between their efforts. The impact of taking shorter showers is not the same as the impact of skipping meat one day a week, which in turn is not the same as the impact of replacing one plane journey with one train journey, for example. This exercise can also be instrumental in preventing a rebound effect. People who save money on energy or on consumption items might be tempted to use that money to consume other things that would annihilate all the carbon savings, such as buying a holiday on the other side of the globe. By the same token, people who do not know the carbon footprint of their different kinds of consumption might be tempted to make some trade-offs to calm their conscience. For example, Berthoû (2013) documents the case of people who eat less meat to compensate for their emissions from a return flight to Thailand, even though one long-distance flight emits more CO_2 than that which can be saved by eating less meat.

Generally speaking, CFCs can help people define their priorities. For example, having one's total carbon footprint displayed as split between energy consumption at home, transport, food, clothes and electronics can help identify low-hanging fruits that can give big reductions with little effort. For many respondents, realizing, for example, that their transport preferences (especially from flights) emit much more CO_2 than energy consumption at home may come as a surprise, because local governments tend to put much emphasis (through information campaigns and various subsidies) on the importance of insulating homes to save on energy bills. Politicians may expect that retrofitting houses might be perceived as less intrusive by citizens, and this can be achieved at a lower political cost than raising taxes on transport. Moreover, big home renovation projects rarely happen – perhaps once or twice in a lifetime (Tjørring and Gausset 2018) – yet they deliver energy savings for a very long period and allow citizens to save significant amounts of money. Yet, looking at the distribution of carbon footprints per sector or activity might help people to prioritize changing their transport habits, for example, since these might account for the lion's share of total private emissions.

The design of some CFCs seems to have been influenced by *Nudge Theory*. Nudging refers to the design and presentation of a variety of choices so that people are likely to choose specific options over others, without economic incentives and without losing the freedom to choose (Thaler and Sunstein 2008). It focusses on how a message can best influence interlocutors, how people are more likely to choose the default option or how displaying goods in a certain way in a supermarket influences people to purchase them, for example. In the context of CFCs, one way of nudging people is to ask them if they intend to change their behaviour, as studies have shown that asking this question is enough to increase the likelihood of people abiding by their intentions. Asking them to make a detailed plan about how they will do this further increases the likelihood that people will do what they intend (ibid.). Some CFCs go one step further and ask respondents to pledge to change their behaviour and to make this pledge public. Pledges are personal commitments to live up to certain goals, and they increase self-awareness, establish explicit personal goals and encourage goal-completing actions. Making a pledge public increases motivation to fulfil the commitment (Abrahamse 2019; Banerjee et al. 2023; Jacobs et al. 2021). Many CFCs ask respondents to define a reduction target or to select deeds and challenges and to report when they have implemented them, thereby keeping track of the emissions they have saved (Pathways, Earth Hero, MyFootprint/SEB, Pawprint or Footprint). Another way of nudging people into continuing to use the app is to reward deeds or climate action with points that can be spent on good causes such as planting trees or providing solar aid in a developing country (TribalData, PawPrint, Greenly or Earth Hero). Meatless also tries to nudge respondents into making long streaks (almost as a series of uninterrupted "good deeds," see Schyberg this volume on an ethnography of carbon in a religious context), while Pawprint or Yayzy try to encourage respondents to repeat good deeds to transform them into "habits" after a given number of repetitions.

Finally, the design of some CFCs seems to have been influenced by *Social Norms Theory*, which posits that people tend to adjust their attitudes and behaviour to what they perceive to be the norm around them (Perkins and Berkowitz 1986; Berkowitz 2004; Perkins 2003). This theory posits that since "perceived norms" often differ from "actual norms," informing people about actual norms can correct "misperceptions" and trigger behavioural adjustment. The success of this kind of approach depends, among other factors, on identification with and proximity to the reference group (Vesely et al. 2022). This normative feedback approach has been used successfully by OPOWER, an American electric utility company that decided in 2009 to inform its customers about how an individual consumer's electric consumption compares to that of the average consumer. The display of this comparison, coupled with a system of differentiated smileys to encourage savings, led to an overall 2% decrease in electric consumption (Allcott 2011). Since the tendency to align one's consumption to social norms can go different ways (it can, for example, encourage people above average to reduce their consumption, but it can also lead people below average to increase their consumption), the use of differentiated smilies in the OPOWER experiment was important to discourage alignment upward. In the same spirit, some CFCs use normative feedback by comparing a respondent's carbon footprint to a national and/or world average (WWF Footprint Calculator, Klima, Earth Hero, Greenly, MyClimate.org or FuturePlanet), to similar social categories in terms of age and income (1.5-degree lifestyles), or to other app users (Pawprint). This kind of feedback might show that someone performing well in terms of the national average may perform poorly in terms of age or income categories, which again may encourage people to make further efforts to align themselves to their social category and make extra efforts to reduce their carbon footprint.

Another aspect of Social Norms Theory is that committing to personal change as part of a group or community might be more motivating to a user than doing it alone. Several CFCs propose the option to create a group with friends or colleagues or to join an existing group (Pathways, Earth Hero, Klima, MyFootprint/SEB or Pawprint). This allows respondents to compare their deeds with those of others, and to – hopefully – be inspired by other peoples' actions. The respondents' good deeds are sometimes ranked to introduce an element of social competition, which may encourage people to make more effort (TribalData or MyFootprint/SEB). But in all of the CFCs reviewed here, collective interactions remain focussed on influencing individual behaviour.

To summarize, in my analysis, most existing individual CFCs rely implicitly on a collection of economic and psychological theories, such as Rational Choice Theory, the Theory of Planned Behaviour, Nudge Theory or Social Norms Theory, all of which posit that people can be influenced through economic incentives, credible information, gentle pushes (rewarding smileys, asking respondents to pledge for change) and by displaying social norms.

Carbon footprint calculators as tools for collective change

Despite their differences, what the CFCs described above have in common is that they all locate change in the choices made by *individuals* who are informed, incentivized and nudged to change their behaviour. Some nudging has a collective dimension, such as informing about actual norms, encouraging good performance with smileys, asking people to make public pledges, or organizing competition with other anonymous users around self-declared deeds and giving points and ranking, but the effectiveness of such approaches seems to be rather small (e.g. a 2% reduction in the case of the OPOWER experiment).

With their strong focus on individual choices, CFCs contribute to taking "carbon as a metric of the human," since "decarbonization implies evaluating how one goes about acting and living" (Whitington 2016: 59; see also the introduction to this volume). By taking self-declared behaviour at face value and by congratulating CFC users on their progress when they push a button on their smartphone, these tools nonetheless run the risk of disembodying the instrument from its target – namely carbon reduction and environmental behaviour (see Bansal and Knox-Hayes 2013 for a comparable argument with carbon-related financial instruments). Moreover, by locating change in personal choices, they place the responsibility for change on individuals, forgetting that choices are collectively shaped by societal infrastructure and political decisions (Paterson and Stripple 2010; Shove 2010; Shove and Trentmann 2019). They might therefore be seen as diverting attention away from the larger societal changes required to solve the global climate crisis.

Somewhere within this classic binary of agency and structure, the research results of the COMPASS project show that practices and behaviour are not just shaped by governmental policies but also by collective action at the grassroots level. In this context, and contrary to the assumptions of change stemming from individual and "rational" choices, CFCs can potentially trigger larger environmental changes through *collective* reflections, choices and actions.

The COMPASS research project designed a survey comparing the carbon footprint of 258 members of green communities. The survey showed that the average carbon footprint of these members is 9 tons, or 28% below the national average of 11.9 tons per capita (Toft et al. 2021). Divided by category, members of green communities are 36% below the average carbon footprint deriving from energy consumption, 27% for transport, 44% for food and 16% for miscellaneous consumption. The results reveal that membership in a green community is the strongest predictor of a low carbon footprint, before income and gender. Moreover, the carbon footprint of members declines with the number of years they spend in the green communities (although the carbon footprint generally increases with age), which demonstrates that even though residents may have had a lower-than-average carbon footprint before joining the green community, membership has also had a separate impact on their carbon footprint (ibid.).

In other words, the results show that collective action leads to lower carbon footprints through grassroots organizations without any specific governmental information campaign, economic incentives or nudges. Instead, carbon footprint reductions derive from the collective design of physical infrastructure (providing collective heating sources, meals, fields, rooms, shared cars, etc.) and social infrastructure to manage it (such as working groups, time banks, communication canals, rules and regulations and social organization to take collective decisions and manage conflicts). These physical and social infrastructures provide members of green communities with a set of default options (e.g. vegetarian communal meals, local district heating, ecological sewage systems, self-produced food or energy), a set of services financed collectively (e.g. shared cars, washing machines, workshops and shared offices) and a set of carbon-saving practices (e.g. borrowing, exchanging or upcycling rather than buying new material goods). Members are free to not use these default options or collective services, but most residents use them because they want to reduce their emissions and because it makes their lives easier and cheaper. Participating in the management of collective infrastructure and in the daily life of the community also leads to a sharing of specialized knowledge, the development of practical skills, a sense of empowerment and higher motivation to live up to the sustainable collective ideals of the community (Gausset 2020; Hansen 2020; Gausset and Hoff 2024; Gausset and Jensen 2024).

What is of special interest for this chapter is that our survey, which was primarily designed as a research project, raised a strong interest in partner green communities and paved the way for collective reflections within them. The carbon footprint performance of the different green communities in the project differed considerably. While the green community with the lowest carbon footprint displayed a 60% reduction in average carbon footprints per capita compared to the national average (Gausset and Jensen 2024), the green communities with the highest carbon footprints were just below the national average. This came as a surprise for one green community, who perceived themselves to be at the forefront of the sustainable transition and had therefore believed that they would perform better than they did. Although they had been at the forefront of the sustainable transition at the time of their establishment, not much had moved on since then, and they had been caught up and even overtaken by municipal services and infrastructures, which had improved tremendously during that period. In another green community, the green ideals that existed in the early days – displayed in the building of high-tech passive houses (ultra-low energy buildings), collective energy production and collective rooms – had not been maintained over time, and there had been no further collective effort to reduce carbon footprints. Our study was used by some of the founders of the community to reopen a debate among their members about the need to be better than average and to do more for the environment. In these cases, the carbon footprint survey triggered a need for a critical re-evaluation of their own societal model and to discuss how it could be revised and improved.

However, the results of our study also came as a surprise in the eco-village that had the lowest carbon footprint (60% below the national average), as they did not perceive themselves to be special or any better than anyone else. Members of this community felt guilty about eating meat, driving cars, flying to holiday destinations and so on. They felt that they had not been making enough effort to live more sustainably. Therefore, they did not realize that they were consuming much less than the national average and that they did this painlessly because of the social and physical infrastructures they had built.

Members of the community have consequently used the results of the research when delivering presentations about their community to the local municipality or in conferences, in an effort to try and spread their model.

In all the cases participating in the study, our carbon footprint survey started a collective reflection on what worked well or less well in their respective green communities. This reflection led to different collective actions, which illustrates that if CFCs can trigger individual behaviour change, they can also lead to collective change. Since Durkheim, it is well known that society is more than the sum of its individual parts. In the present context, this means that collective action can trigger more societal change than individual action (see also Gausset 2013; Tjørring and Gausset 2015; Lex and Gausset 2022; Nielsen-Englyst and Gausset 2024).

Conclusion

CFCs are tools that measure carbon emissions associated with activities, production or consumption. When targeting individuals, they are designed to evaluate private carbon footprints, either with the goal of selling carbon offsets or with the goal of triggering individual behaviour change by helping users measure the origin and amount of their CO_2 emissions, defining targets and priorities for emission reductions and monitoring progress to reach these targets.

Although CFCs take self-declared behaviour at face value, their actual impact on environmental behaviour remains largely undocumented. Moreover, by locating change in personal choices, they place the responsibility for change on individuals, when choices are also collectively shaped by infrastructure, policymaking and local communities, as I have argued here.

In my analysis, most existing CFCs seem to rely implicitly on economic and/or psychological approaches, assuming that individual behaviour can be changed through providing the right information, communication, incentives and nudges, which are used to variable degrees in the different CFCs. They thus reinforce the individualization of responsibility criticized by several of the scholars referenced above. Yet, I have shown that CFCs also have the potential to be used in a more anthropological or sociological approach to inform and trigger *collective* change and *collective* action around common sustainability goals, which can – according to our study – deliver a much greater reduction in carbon footprints than private and individual action.

Notes

1 Some CFCs have been used in experimental pilot projects that provide citizens with an equal personal carbon allowance (PCA). PCAs are government-issued permits that allow an individual or a company to release a certain quota of CO_2 emissions (Fawcett and Parag 2010). At an individual level, each time someone buys fuel for transport, for house heating, or (in more ambitious schemes) food and manufactured items and services, the related emissions are deducted from a personal budget.

People who have exhausted their budget are then able to purchase additional credits on the personal carbon market from those with excess to sell (Fuso et al. 2021). Personal carbon allowances are highly complex projects and can be costly to implement (ibid.), but there are some pilot projects that explore this option, for example, in Lahti, Finland (Kuokkanen et al. 2020; Elisa et al. 2021; Ville et al., 2022).
2 There is a fast-growing market of offsetting carbon emissions (measured through a CFC), both for enterprises and for individuals (Forest Trends' Ecosystem Marketplace 2021). Carbon emissions can be offset though carbon sink projects, such as reforestation projects in which the released emissions are captured and stored by planting trees. They can also be offset by purchasing carbon credits on a Cap-and-Trade system, such as the European Union's Emissions Trading System (ETS), in which power plants, oil refineries, airlines and other industrial facilities are given an emissions limit ("cap"), which cannot be exceeded over a certain period. Companies can purchase or sell allowances to stay within their emissions limit.
3 Eco-villages are communities of people that seek to live more sustainable lives. They typically consist of people living in private houses and sharing collective buildings, fields and activities with other residents, in order to reduce their carbon footprint and environmental impact. Green neighbourhoods are communities of neighbours who voluntarily decide to meet locally, discuss what kind of sustainable interests they share, and act together to live more sustainably. Organic food cooperatives consist of people joining forces to buy organic vegetables directly from local producers and distributing them among members. Organic vegetables are typically more local, fresher, more diverse and more seasonal, and therefore have a lower environmental impact than vegetables found in supermarkets – especially when they are grown following the principles of regenerative agriculture.
4 We did not take emissions embedded in their production into account.
5 CFCs displaying Scope 3 emissions from societal infrastructure might help to discourage behaviour change, since a respondent's carbon footprint might always end up being above sustainable levels of emissions, regardless of the efforts they make to reduce it (see also Franz and Papyrakis 2011).

References

Abrahamse, Wokje. 2019. *Encouraging Pro-Environmental Behaviour: What Works, What Doesn't, and Why*. London: Academic Press.

Ajzen, Icek. 1991. "The Theory of Planned Behavior." *Organizational Behavior and Human Decision Processes* 50, no. 2 (December): 179–211. https://doi.org/10.1016/0749-5978(91)90020-T

Allcott, Hunt. 2011. "Social Norms and Energy Conservation." *Journal of Public Economics* 95, no. 9–10 (October): 1082–1095. https://doi.org/10.1016/j.jpubeco.2011.03.003

Andersson, David. 2020. "A Novel Approach to Calculate Individuals' Carbon Footprints Using Financial Transaction Data – App Development and Design." *Journal of Cleaner Production* 256 (May): 120396. https://doi.org/10.1016/j.jclepro.2020.120396

Banerjee, Sanchayan, Matteo M. Galizzi, Peter John, and Susana Mourato. 2023. "Sustainable Dietary Choices Improved by Reflection Before a Nudge in an Online Experiment." *Nat Sustain* 6 (October): 1632–1642. https://doi.org/10.1038/s41893-023-01235-0

Bansal, Pratima, and Janelle Knox-Hayes. 2013. "The Time and Space of Materiality in Organizations and the Natural Environment." *Organization & Environment* 26, no. 1 (January): 61–82. https://doi.org/10.1177/1086026612475069

Barendregt, Wolmet, Aksel Biørn-Hansen, and David Andersson. 2020. "Users' Experiences with the Use of Transaction Data to Estimate Consumption-Based Emissions in a Carbon Calculator." *Sustainability* 12, no. 18 (September): 7777. https://doi.org/10.3390/su12187777

Berkowitz, Alan David. 2004. "An Overview of the Social Norms Approach". In *Changing the Culture of College Drinking: A Socially Situated Prevention Campaign*, edited by L. Lederman, L. Stewart, F. Goodhart and L. Laitman. New York: Hampton Press.

Berthoû, Sara Kristine Gløjmar. 2013. "The Everyday Practices of Pro-Environmental Practices". *The Journal of Transdisciplinary Environmental Studies* 12 no. 1: 53–68.

Birnik, Andreas. 2013. "An Evidence-Based Assessment of Online Carbon Calculators." *International Journal of Greenhouse Gas Control* 17 (September): 280–293. https://doi.org/10.1016/j.ijggc.2013.05.013

Bjørn-Hansen, Aksel, Cecilia Katzeff and Elina Eriksson. 2022. "Exploring the Use of a Carbon Footprint Calculator Challenging Everyday Habits." Paper presented at the *Nordic Human-Computer Interaction Conference (NordiCHI)*, Aarhus, Denmark, October 8–12, 2022, pp. 1–10. New York: Association for Computing Machinery. https://doi.org/10.1145/3546155.3546668

Burgui-Burgui, Mario, and Emilio Chuvieco. 2020. "Beyond Carbon Footprint Calculators. New Approaches for Linking Consumer Behaviour and Climate Action." *Sustainability* 12, no. 16 (August): 6529. https://doi.org/10.3390/su12166529

Collins, Andrea, Alessandro Galli, Nicoletta Patrizi, and Federico Maria Pulselli. 2018. "Learning and Teaching Sustainability: The Contribution of Ecological Footprint Calculators." *Journal of Cleaner Production* 174 (February): 1000–1010. https://doi.org/10.1016/j.jclepro.2017.11.024

Collins, Andrea, Alessandro Galli, Tara Hipwood, and Adeline Murthy. 2020. "Living within a One Planet Reality: The Contribution of Personal Footprint Calculators." *Environmental Research Letters* 15, no. 2 (February): 025008. http://doi.org/10.1088/1748-9326/ab5f96

Dreijerink, Lieke, and Geerte Paradies. 2020. *How to Reduce Individual Environmental Impact? A Literature Review into the Effects and Behavioral Change Potential of Carbon Footprint Calculators*. Amsterdam: TNO.

Edstrand, Emma. 2016. "Making the Invisible Visible: How Students Make Use of Carbon Footprint Calculator in Environmental Education." *Learning, Media and Technology* 41, no. 2 (May): 416–436. https://doi.org/10.1080/17439884.2015.1032976

Enlund, Jacob, David Andersson, and Frederik Carlsson. 2023. "Individual Carbon Footprint Reduction: Evidence from Pro-environmental Users of a Carbon Calculator." *Environmental and Resource Economics* 86 (August): 433–467. https://doi.org/10.1007/s10640-023-00800-7

Fawcett, Tina, and Yael Parag. 2010. "An Introduction to Personal Carbon Trading." *Climate Policy* 10, no. 4 (June): 329–338. https://doi.org/10.3763/cpol.2010.0649

Fishbein, Martin. 1963. "An Investigation of Relationships between Beliefs about an Object and the Attitude toward that Object." *Human Relations* 16, no. 3 (August): 233–240. https://doi.org/10.1177/001872676301600302

Fishbein, Martin. 1967. "Attitude and the Prediction of Behaviour". In *Readings in Attitude Theory and Measurement*, edited by M. Fishbein. New York: Wiley.

Fishbein, Martin. 1968. "An Investigation of Relationships Between Beliefs about an Object and the Attitude Towards That Object." *Human Relationships* 16, no. 3 (August): 233–239. https://doi.org/10.1177/001872676301600302

Fishbein, Martin, and Icek Ajzen. 1972. *Beliefs, Attitudes, Intentions and Behaviour: An Introduction to Theory and Research*. Reading Mass. Boston: Addison-Wesley.

Fishbein, Martin, and Icek Ajzen. 1974. "Attitudes Towards Objects as Predictors of Single and Multiple Behavioural Criteria." *Psychological Review* 81, no. 1: 29–74. https://psycnet.apa.org/doi/10.1037/h0035872

Forest Trends' Ecosystem Marketplace. 2021. *'Market in Motion', State of Voluntary Carbon Markets 2021, Installment 1*. Washington DC: Forest Trends Association.

Franz, Jennifer, and Elissaios Papyrakis. 2011. "Online Calculators of Ecological Footprint: Do They Promote or Dissuade Sustainable Behaviour?" *Sustainable Development* 19, no. 6 (November): 391–401. https://doi.org/10.1002/sd.446

Fuso Nerini, Francesco, Tina Fawcett, Yael Parag, and Paul Ekins. 2021. "Personal Carbon Allowances Revisited." *Nat Sustain* 4 (August): 1025–1031. https://doi.org/10.1038/s41893-021-00756-w

Gausset, Quentin. 2013. "Comparing Different Approaches to Environmental Behavioural Change: A Review of Ten Case Studies from Denmark." *The Journal of Transdisciplinary Environmental Studies* 12 no. 1: 30–40.

Gausset, Quentin. 2020. "Stronger Together. How Danish Environmental Communities Influence Behavioural and Societal Changes." In *The Role of Non-State Actors in the Green Transition: Building a Sustainable Future*, edited by J. Hoff, Quentin Gausset, and Simon Lex, 52–70. London: Routledge. https://doi.org/10.4324/9780429280399

Gausset, Quentin, and Jens Hoff. 2024. "Fællesskab – Et overset element i den grønne omstilling." In *Grundbog om grøn samfundsteori*, edited by Stine Krøijer and Lars Tønder. Copenhagen: Gyldendal.

Gausset, Quentin, and Pia Duus Jensen. 2024. "Living Sustainably in a Danish Eco-community: How Social and Physical Infrastructures Affect Carbon Footprints." *Climate Action* 33, no. 3 (May). DOI: 10.1038/s44168-024-00113-5

Gram-Hanssen, Kirsten, and Toke Haunstrup Christensen. 2012. "Carbon Calculators as a Tool for a Low-Carbon Everyday Life?" *Sustainability: Science, Practice and Policy* 8, no. 2 (October): 19–30, https://doi.org/10.1080/15487733.2012.11908093

Hansen, A.H. 2020. "It Has to Be Reasonable: Pragmatic Ways of Living Sustainably in Danish Ecocommunities." In *The Role of Non-State Actors in the Green Transition: Building a Sustainable Future*, edited by Jens Hoff, Quentin Gausset, and Simon Lex, 34–51. London: Routledge. https://doi.org/10.4324/9780429280399

Jacobs, Tyler P., Lauren L. Gottschalk, Mitchell Dandignac, and Allen R. McConnell. 2021. "Making Pledges More Powerful: Effects on Pro-Environmental Beliefs and Conservation Behavior." *Sustainability* 13, no. 17 (September): 9894. https://doi.org/10.3390/su13179894

Jones, Christopher M., and Daniel M. Kammen. 2011. "Quantifying Carbon Footprint Reduction Opportunities for U.S. Households and Communities." *Environmental Science & Technology* 45, no. 9 (March): 4088–4095. https://doi.org/10.1021/es102221h

Knox, Hannah. 2020. *Thinking Like a Climate Governing a City in Times of Environmental Change*. Durham: Duke University Press.

Kok, Anne Linda, and Wolmet Barendregt. 2021. "Understanding the Adoption, Use, and Effects of Ecological Footprint Calculators among Dutch Citizens." *Journal of Cleaner Production* 326 (December): 129341, https://doi.org/10.1016/j.jclepro.2021.129341

Kuokkanen, Anna, Markus Sihvonen, Ville Uusitalo, Anna Huttunen, Tuuli Ronkainen, and Helena Kahiluoto. 2020. "A Proposal for a Novel Urban Mobility Policy: Personal Carbon Trade Experiment in Lahti City." *Utilities Policy* 62 (February): 100997. https://doi.org/10.1016/j.jup.2019.100997

Lex, S., and Quentin Gausset. 2022. "Fællesskab som drivkraft for den grønne omstilling." *Økonomi og Politik* 2, no. 2 (October): 51–60. https://doi.org/10.7146/okonomiogpolitik.v95i2.134165

Lippert, Ingmar. 2015. "Environment as Datascape: Enacting Emission Realities in Corporate Carbon Accounting." *Geoforum* 66 (November): 125–135, https://doi.org/10/wx8

Lippert, Ingmar. 2018. "On Not Muddling Lunches and Flights: Narrating a Number, Qualculation, and Ontologising Troubles." *Science and Technology Studies* 31, no. 4 (December): 52–74, https://doi.org/10.23987/sts.66209

Mackenzie, Donald. 2009. "Making Things the Same: Gases, Emission Rights and the Politics of Carbon Markets." *Accounting, Organizations and Society* 34, no. 3–4 (Spring): 440–455. https://doi.org/10.1016/j.aos.2008.02.004

Mulrow, John, Katherine Machaj, Joshua Deanes, and Sybil Derrible. 2019. "The State of Carbon Footprint Calculators: An Evaluation of Calculator Design and User Interaction Features." *Sustainable Production and Consumption* 18 (April): 33–40. https://doi.org/10.1016/j.spc.2018.12.001

Nielsen-Englyst, Camilla, and Quentin Gausset. 2024. "From Countercultural Ecovillages to Mainstream Green Neighbourhoods - A View on Current Trends in Denmark." https://doi.org/10.21203/rs.3.rs-2640366/v1

Pandey, Divya, Madhoolika Agrawal, and Jai Shanker Pandey. 2011. "Carbon Footprint: Current Methods of Estimation." *Environ Monit Assess* 178 (September): 135–160. https://doi.org/10.1007/s10661-010-1678-y

Paterson, Matthew, and Johannes Stripple. 2010. "My Space: Governing Individuals' Carbon Emissions." *Environment and Planning D: Society and Space* 28, no. 2 (January): 341–362. https://doi.org/10.1068/d4109

Perkins, H. Wesley. 2003. "The Emergence and Evolution of the Social Norms Approach to Substance Abuse Prevention". In *The Social Norms Approach to Preventing School and College Age Substance Abuse: A Handbook for Educators, Counselors, and Clinicians*, edited by H. Wesley Perkins. San Francisco: Jossey-Bass.

Perkins, H. Wesley and Alan David Berkowitz. 1986. "Perceiving the Community Norms of Alcohol Use Among Students: Some Research Implications for Campus Alcohol Education Programming." *International Journal of the Addictions* 21, no. 9–10 (Fall): 961–976. DOI: 10.3109/10826088609077249

Salo, Marja, Ari Nissinen, Raimo Lilja, Emilia Olkanen, Mia O'Neill, and Martina Uotinen. 2016. "Tailored Advice and Services to Enhance Sustainable Household Consumption in Finland." *Journal of Cleaner Production* 121 (May): 200–207. https://doi.org/10.1016/j.jclepro.2016.01.092

Salo, Marja, Marja K. Mattinen-Yuryev, and Ari Nissinen. 2019. "Opportunities and Limitations of Carbon Footprint Calculators to Steer Sustainable Household Consumption – Analysis of Nordic Calculator Features." *Journal of Cleaner Production* 207 (January): 658–666. https://doi.org/10.1016/j.jclepro.2018.10.035

Scrucca, Flavio, Grazia Barberio, Valentina Fantin, Pier Luigi Porta, and Marco Barbanera. 2021. "Carbon Footprint: Concept, Methodology and Calculation." In *Carbon Footprint Case Studies. Environmental Footprints and Eco-design of Products and Processes*, edited by Subramanian Senthilkannan Muthu. Singapore: Springer. https://doi.org/10.1007/978-981-15-9577-6_1

Shove, Elizabeth. 2010. "Beyond the ABC: Climate Change Policy and Theories of Social Change." *Environment and Planning A: Economy and Space* 42, no. 6 (June): 1273–1285. https://doi.org/10.1068/a42282

Shove, Elizabeth, and Frank Trentmann. 2019. *Infrastructures in Practice: The Dynamics of Demand in Networked Societies*. London: Routledge. https://doi.org/10.4324/9781351106177

Strobel, Bjarne W., Anders Christian Erichsen, and Quentin Gausset. 2016. "The conundrum of calculating carbon footprints." In *Community Governance and Citizen-Driven Initiatives*, edited by Jens Hoff and Quentin Gausset, pp. 7–27. London: Routledge. DOI: 10.4324/9781315700298-2

Thaler, Richard H., and Cass R. Sunstein. 2008. *Nudge: Improving Decisions About Health, Wealth and Happiness*. New Haven: Yale University Press.

Tjørring, Lise, and Quentin Gausset. 2015. "Energy Renovation Models in Private Households in Denmark." In *Community Governance and Citizen-Driven Initiatives in Climate Change Mitigation*, edited by Jens Hoff and Quentin Gausset, pp. 89–106. London: Routledge. DOI: 10.4324/9781315700298-6

Tjørring, Lise, and Quentin Gausset. 2018. "Drivers for Retrofit: A Sociocultural Approach to Houses and Inhabitants." *Building Research and Information* 47, no. 4 (February): 394–403. https://doi.org/10.1080/09613218.2018.1423722

Toft, M., Quentin Gausset, Jens Hoff, and Simon Lex. 2021. "Together We Green: Measuring the Community Effect on Environmental Behaviour." Copenhagen University, Denmark. Draft. [Unpublished manuscript].

Uusitalo, Elisa, Anna Kuokkanen, Ville Uusitalo, Tuuli von Wright, and Anna Huttunen. 2021. "Personal Carbon Trading in Mobility May Have Positive Distributional Effects." *Case Studies on Transport Policy* 9, no. 1 (March): 315–323. https://doi.org/10.1016/j.cstp.2021.01.009

Uusitalo, Ville, Anna Huttunen, Elisa Kareinen, Tuuli von Wright, M. Valjakka, Atte Pitkänen, and Jarkko Levänen. 2022. "Using Personal Carbon Trading to Reduce Mobility Emissions: A Pilot in the Finnish City of Lahti" *Transport Policy* 126 (September): 177–187. https://doi.org/10.1016/j.tranpol.2022.07.022

Vesely, Stepan, Christian A. Klöchner, Giuseppe Carrus, Lorenza Tiberio, Federica Caffaro, Mehmet Efe Biresselioglu, Andrea C. Kollmann, and Anca C. Sinea. 2022. "Norms, Prices, and Commitment: A Comprehensive Overview of Field Experiments in the Energy Domain and Treatment Effect Moderators." *Frontiers in Psychology* 13 (November): 967318. https://doi.org/10.3389/fpsyg.2022.967318

West, Sarah E., Anne Owen, Katarina Axelsson, and Chris D. West. 2016. "Evaluating the Use of a Carbon Footprint Calculator: Communicating Impacts of Consumption at Household Level and Exploring Mitigation Options." *Journal of Industrial Ecology* 20, no. 3 (November): 396–409. https://doi.org/10.1111/jiec.12372

Whitington, Jerome. 2016. "Carbon as a Metric of the Human." *PoLAR* 39, no. 1 (May): 46–63. https://doi.org/10.1111/plar.12130

3 Configuring the carbon farmer

Emerging practices of carbon accounting and biochar engagements in Danish agriculture

Inge-Merete Hougaard

Introduction: the emerging carbon farmer

Michael walks the perimeter of the field where biochar was applied last year.[1] At one end of the field, an irrigation machine pumps rhythmically, spraying water in circles, leaving a trace of coloured drizzle in the air. Michael says that it hasn't rained for months. We walk across the field looking for biochar. "There is not much to see," he says. "If you're lucky, you can find a small piece." He scratches the surface soil with his clogs and bends down: "Here is a bit." He picks up a small black pellet the size of a fingernail and places it in the palm of his hand. It is an unremarkable indication that biochar has been applied to the field, where small green leaves have started to find their way into the spring sunlight. "It is spring barley," Michael says to help my untrained eye identify the crop. Due to biochar's time lag in effect, we cannot tell the difference between the plants that have received biochar and those that have not. "Last year we had maize here," Michael adds, before we walk back to the farm with his two dogs circling around our feet.

Michael and his wife Karen are part of a field experiment aimed at testing different methods for applying biochar to agricultural soils. While there have been earlier attempts to introduce biochar as a method for soil improvement in Denmark, these never got beyond test sites of researchers, niche agriculturalists and entrepreneurial estates. However, as biochar has been identified as a method for carbon dioxide removal (CDR) by the international climate research community (see e.g. Minx et al. 2018), it has become the subject of renewed attention in Danish agricultural and climate policy. In recent years, governments and corporations have positioned CDR as a key strategy to reach net-zero emission goals (Battersby et al. 2022; Buylova et al. 2021; Schenuit et al. 2021). In modelled pathways, CDR methods are envisioned to balance out so-called hard-to-abate emissions in net-zero scenarios and compensate for a potential carbon budget overshoot towards the end of the century (Lund et al. 2023; Fuss et al. 2018). Produced through the process of pyrolysis, biochar is considered one of various CDR methods since it consists of biomass-based carbon that is stabilized in a charcoal-like form (see elaboration below).[2] Other CDR methods include afforestation/reforestation, bioenergy with carbon capture

DOI: 10.4324/9781003478669-4

and storage (BECCS), direct air carbon capture and storage (DACCS), enhanced weathering, and ocean fertilization (Minx et al. 2018). These practice-technologies[3] differ in terms of commercial maturity, expected economic costs, envisioned mitigation potential and anticipated environmental impacts (Minx et al. 2018; Creutzig et al. 2021). However, biochar is considered to be more cost-effective and has less land, water and energy requirements than other CDR (hereafter "carbon removal") methods (Pratt and Moran 2010; Smith 2016). Yet, none of these methods currently work at the scales envisioned in modelled scenarios, and along with justice concerns about who should be credited for the carbon removed (Armstrong and McLaren 2022), scholars warn that the promise of future carbon removal may result in mitigation deterrence; that is, diminished or delayed efforts to reduce emissions due to expectations of future climate measures (Markusson, McLaren and Tyfield 2018; Carton et al. 2023). Importantly, reductions and removals are not equivalent but have very different social, economic and ecological implications for people and environments across different geographies and positions of power (McLaren et al. 2019; Carton, Lund and Dooley 2021; Dalsgaard 2013).

This chapter explores new social configurations that are emerging with the introduction of biochar in Danish climate policy. As a country considered by itself and others as a climate frontrunner (Dyrhauge 2021; Jänicke and Wurzel 2019), Denmark has embraced carbon removal in its climate agenda. The parliament has set an emission reduction goal of 70% by 2030 and a net-zero goal by 2050 (Danish Government 2019). In the first Climate Programme, which was to illustrate how the 2030 goal will be reached, 7–12 of the required 20 Mt CO_2e reductions[4] involved methods for carbon removal[5] (DMCEU 2020). Of the 7–12 Mt CO_2e, it was imagined that biochar would deliver 2 Mt CO_2e, corresponding to 10% of the needed national reductions.[6] This indicates that reductions and removals are conflated (instead being counted separately) in climate targets, which implies complicated accounting practices that increase the risk of mitigation deterrence (McLaren et al. 2019).

The emergence of carbon removal and net-zero goals, thus, invites us to revisit the idea of carbon accounting. Carbon accounting involves determining an entity's current emissions in order to reduce these – or compensate for them through external offsets. It has mainly been studied in contexts of green corporate performance (Lippert 2015; Lovell and Mackenzie 2011; MacKenzie 2009), individual carbon footprints (Bruckermann 2022; Girvan 2018; Lovell, Bulkeley and Liverman 2009) or North-South compensation mechanisms in global climate politics (Ervine 2013; Paladino and Fiske 2017; Bryant, Dabhi and Böhm 2015). Carbon accounting and offsetting can thus bring about two forms of "carbon subjects." First, they bring about carbon-emitting individuals or companies who are deemed "guilty" by themselves or others and must account for their emissions and reduce (or compensate for) these (Bruckermann 2022; Girvan 2018; Lovell, Bulkeley and Liverman 2009; Lippert 2015). And, second, they bring about offsetting providers who must adhere to neoliberal management regimes of measurements, standards and certification schemes to participate in the carbon economy (Asiyanbi, Ogar and Akintoye 2019; Paladino and Fiske 2017; Fletcher 2010). Contemporary carbon removal hence builds on a long history of carbon sinks and sequestration (mainly) in the Global South that offsets emissions (mainly) in the Global North (Carton et al. 2020; Gutiérrez 2016). Yet, after decades of critique of climate colonialism (Navarro 2022; Zografos and Robbins 2020; Bachram 2004), carbon removal (and offsetting) activities are now also being envisioned to be implemented in the Global North. While international offsetting has been widely criticized for problems of ethics, fairness, additionality, leakage, temporality, double counting and reproducing unequal global relations (Dalsgaard 2013; Ervine 2013; Gutiérrez 2011; Lohmann 2008, 2009; Lovell, Bulkeley and Liverman 2009; MacKenzie 2009), the Global North's domestication of carbon removal activities may address some of these challenges. However, as market, labour and power relations are reconfigured, other problems may (re-)appear. Moreover, since new actors (such as farmers) are both required to reduce their emissions (Bruckermann 2022; Girvan 2018; Lovell, Bulkeley and Liverman 2009; Lippert 2015) and envisioned to engage in carbon removal activities (see e.g. Stanley 2024), new carbon subjects are emerging.

In Denmark, biochar is attributed to the agricultural sector,[7] a key area for Danish climate policy. While agriculture is considered to account for 22% of global emissions (IPCC 2022), in Denmark the sector is responsible for 34% of national emissions – a proportion expected to increase in coming years as other sectors decarbonize further (DEA 2023). Agriculture occupies a relatively privileged position in Denmark, in part because one of the two main political parties, the Liberal Party (Venstre), has roots in the peasant movement and is still considered to represent farmer interests (Juul Christiansen and Flemming 2020). However, while the agricultural sector promotes itself as having "one of the most climate-friendly productions in the world" (DAFC 2022, 5), it has in recent years been positioned by the surrounding society as a "climate culprit" given its environmental and climate impact (Kramer 2022; Hastrup, Brichet and Nielsen 2022). While the sector was initially exempted from a national CO_2 tax (Danish Government 2022), the prospects of such a tax have put pressure on farmers to increase their climate mitigation efforts. Biochar is thus seen as a welcomed climate mitigation option by the sector. It is envisioned to account for 25–33% of sectoral reductions and was for long the largest single initiative in the sector in terms of contributing towards the 2030 reduction goal (Danish Government 2021). Biochar is considered compatible with dominant production practices (Kon Kam King et al. 2018) and matches existing narratives of Danish farmers being at the forefront of technological innovation. Further, it speaks to the idea of "carbon farming," where farmers are envisioned to use land-based methods to capture and store CO_2 (Dumbrell, Kragt and Gibson 2016; Okyere and Kornher 2023; Paul et al. 2023). With initiatives like the EU's anticipated carbon farming scheme (European Commission 2021), I suggest that the EU social contract, where farmers are supported in turn of "providing food for the people" (European Commission 2023), is changing: farmers are in the future not only receiving support for being food farmers; they also receive it for being *carbon farmers*, managing the land to provide climate mitigation through carbon sequestration.

In this chapter, I explore how Danish farmers are configured as carbon farmers through the introduction of biochar as a carbon removal method. In addition to the concept of "carbon subjects" (Lovell, Bulkeley and Liverman 2009; Girvan 2018; Bruckermann 2022), I use that of "user configuration" (Oudshoorn and Pinch 2003; Woolgar 1990) to explore how farmers imagine and are imagined engaging with biochar as a practice-technology by researchers, pyrolysis entrepreneurs, agricultural consultants and agricultural interest organizations. According to Woolgar (1990), user configuration involves defining or negotiating who a user might be and setting boundaries for what kind of actions a user can (and cannot) perform. Oudshoorn and Pinch (2003) highlight that this is not an automatic process but something that requires deliberate work; inventors have to figure out who the users might be and how they will engage with technology. Yet, the envisioned user identities may not be explicitly articulated but negotiated in the social space between different actors along the production, distribution and marketing chain (Woolgar 1990). Thus, user configuration is an ongoing process, where users and technologies are continuously co-constructed (Sánchez-Criado et al. 2014; Oudshoorn and Pinch 2003).

I suggest that the introduction of biochar configures Danish farmers as carbon subjects in two ways: on the one hand, they are cast as "guilty" individuals (or companies[8]) compelled to address their carbon emissions (Bruckermann 2022; Girvan 2018; Lovell, Bulkeley and Liverman 2009; Lippert 2015). On the other hand, they are positioned as potential carbon farmers and carbon removal providers, subject to neoliberal management schemes of measurements, standards and certifications (Asiyanbi, Ogar and Akintoye 2019; Paladino and Fiske 2017; Fletcher 2010). However, I argue that the move from "climate culprit" (Kramer 2022) to carbon farmer effectively requires farmers to engage in both new farming and carbon accounting practices. Yet, I find that the configuration of new farming practices is often implicit or postponed, and the dominant focus is on "getting the technology right" (i.e. the pyrolysis process). The mundane practices of producing biomass for the pyrolysis process and applying the resulting biochar to soils are neglected or taken for granted by non-farmer actors. This results in a disconnect between the farmers' and other actors' expectations about how farmers should engage with biochar. Furthermore, the focus on technology leaves out important debates about the distribution of climate and economic benefits, resulting in diverging ideas regarding who should be credited with the resulting carbon removal and associated economic gains. This risks resulting in double claiming (Kreibich 2024; Schneider and La Hoz Theuer 2019), where different actors claim the same removals, making climate mitigation efforts seem larger than they are.[9] Overall, I contend that the case illustrates what Günel (2019) calls a "technical adjustment," where actors in and around the agricultural sector demonstrate willingness to establish new complex technical systems in order to avoid larger changes in existing production models.

This chapter draws on 23 semi-structured interviews with and nine field visits to actors in and around the Danish biochar sector, including farmers, pyrolysis entrepreneurs, researchers, agricultural consultants and agricultural interest organizations. Interviews and field visits took place between October 2020 and September 2022. In this period, I also conducted participant observation at six events: two sector conferences, two webinars and two network meetings. To some degree I was "studying up" and "studying sideways" (Nader 1972), as I engaged with actors in relatively privileged positions in organizations that seek to influence climate governance, but also with actors with less strategic approaches to biochar. I analysed transcribed interviews and fieldnotes and detected central themes in an iterative dialogue between theoretical frameworks and my data. Translations from Danish to English are my own.

Following this introduction, I briefly sketch how biochar entered the climate mitigation agenda and the Danish context. I then describe how different actors in the agricultural sector imagine farmers should engage with biochar. By way of conclusion, I reflect on the changes and continuities that counting carbon means in the Danish agricultural sector.

Anticipating new farming and carbon accounting practices

Biochar is produced through pyrolysis, the incomplete combustion of biomass at high temperatures (over 300°C) under oxygen-deprived conditions. Apart from biochar, the pyrolysis process produces different proportions of bio-oil and pyrogas depending on the input material and the temperatures employed (Schmidt et al. 2019; Otte and Vik 2017). If biomass input material is combusted under normal circumstances or left to decompose, it releases the CO_2 it has absorbed through photosynthesis, which is why biomass combustion is considered CO_2-neutral in international carbon accounting. However, the pyrolysis process ensures that carbon is transformed into solid biochar material, which is deemed stable for up to 1,000 years (Bis et al. 2018). By "storing" the carbon that would otherwise have been emitted, biochar is considered a form of carbon removal. When applied to agricultural land, biochar is considered to improve soil structure, water retainment and nutrient uptake (Elsgaard et al. 2022; Otte and Vik 2017; Smith 2016).

Biochar was first documented in research reports in the 1870s as a soil-improving practice employed in the Amazon 500–2,500 years ago (Fairhead, Leach and Kojo 2012; Bezerra 2015). In the 1990s, it entered the sphere of international environmental governance as a technique for improving agricultural production and fighting climate change (Soentgen et al. 2017). Initially known as Terra Preta do Índio or Amazonian Dark Earths (ADE),[10] it was increasingly commodified and commercialized as biochar, and its climate mitigation potential attained growing importance (Leach, Fairhead and Fraser 2012; Soentgen et al. 2017; Bezerra et al. 2019). At first, the production and use of biochar was aimed at low-income countries, where it was envisioned to generate carbon offsets by enrolling small-scale farmers into climate compensation schemes (Leach, Fairhead and Fraser 2012; Soentgen et al. 2017; Hansson et al. 2021; Kon Kam King et al. 2018). In recent years, it has emerged as a tool for climate mitigation also in Northern Europe (Azzi, Karltun and Sundberg 2021; Bis et al. 2018; Hagenbo et al. 2022; Tammeorg et al. 2021), particularly in the context of net-zero targets and carbon removal.

This is also the case in Denmark, where biochar, with impetus from the Paris Agreement and the IPCC's 1.5°C report (2018), was placed on the climate agenda as a carbon removal method. Earlier attempts at introducing biochar had not been successful – its increased crop production and soil-improving effects had not been demonstrated in Denmark due to agricultural fields already being highly fertilized (Elsgaard et al. 2022).[11] Despite its big role in national climate policy, at the time of research, biochar had not been employed in Denmark outside of field experiments and test sites. Estimates indicate that depending on the input material, soil type, carbon content and frequency of application, biochar would need to be applied to 10–50% of current Danish agricultural lands each year to deliver the anticipated 2 Mt CO_2e, while other scenarios claim that it would necessitate up to five times the current agricultural area (Thomsen, Karlsson and Kamp 2023; Elsgaard et al., 2022). These figures indicate that national climate policy has not considered the practical implementation of biochar and that many (if not all) Danish farmers would have to take on new biochar-related agricultural practices if biochar were to deliver removals on the envisioned scale (Hougaard 2024). As illustrated below, this change in agricultural practices was largely ignored in policy and public debate.

Further, the introduction of biochar as a carbon removal method anticipates that farmers engage in new accounting practices. As carbon removal providers, farmers would need to document the carbon stored in their biochar and provide accounts of their farm's emissions if they wish to subtract it from these. As elaborated below, farmers are increasingly compelled to engage in carbon accounting due to the anticipated CO_2 tax and new demands from financial institutions and agrifood industries. Thus, I suggest that the introduction of biochar as a carbon removal method anticipates that farmers adopt both new farm practices and new accounting practices, which are so far only vaguely imagined.

A double carbon subject

The reception at the agricultural interest organization is busy: people coming in and out of the elevators, phones ringing almost incessantly, receptionists answering calls and attending guests, and guests waiting in the comfortable lobby chairs. Behind the chairs, a large painting of a field being harvested decorates the wall. On the opposite wall, three screens list current meetings, their starting times and places. I am here to talk to Morten, a policy consultant from the organization, about their visions for biochar in Danish climate policy.

Morten leads me through busy office corridors into a vacant meeting room. He explains that the organization's interest in biochar started when they launched their mission to become climate-neutral by 2050. He says that while they could demonstrate that CO_2 emissions in the agricultural sector had decreased since the 1990s, it was increasingly more difficult to find ways to make further reductions in the future. "There were the easy things, those we knew (…) But we were looking for some groundbreaking things. (…) And then in this process, these [pyrolysis] people appeared…" He frames the organization's emerging interest in biochar in the context of the EU carbon farming agenda, where the land sector is considered important for climate neutrality due to its ability to "capture CO_2 from the atmosphere" (European Commission 2021). "And this is where it becomes really interesting," Morten says, and compares biochar to conservation agriculture and carbon capture and storage (CCS): "If you look at biochar. You can't deny that it can store [carbon] and it can stay there for up to 1,000 years. (…) So that was super interesting for us to look into."

Morten explains that the organization considers biochar as important for meeting the national climate target and that they carried the idea of biochar onto the national climate agenda. He says that farmers see biochar as "something that can save them," implicitly referring to the anticipated CO_2 tax and the discourse of farmers as climate culprits. Morten says that farmers are "very interested" in engaging with biochar. "It is almost always part of the programme when we have a congress. (…) They want to hear about the financial model and what they can get out of it." The challenge, he adds, is how to create an economic model, but here the EU carbon farming agenda offers opportunities: "It actually mentions biochar in some of [the material]." Though his organization "is not a big fan of a national CO_2 tax," Morten acknowledges that it probably will be implemented and thereby "establish a price on carbon" in the agricultural sector. He anticipates a similar mechanism at the EU level: "And that creates a different picture (…) with another platform to get an economic model for pyrolysis." Hence, on the one hand, Morten indicates that Danish society positions farmers as "guilty" carbon subjects who need to manage their greenhouse gas (GHG) emissions, and, on the other, he points to the opportunities for farmers to become carbon removal providers who can benefit from the carbon economy.

The double positioning of farmers as climate culprits and potential beneficiaries of biochar is also recognized among regional agricultural consultants. Kai, who advises farmers on their production practices on a daily basis, explains that his entry point to working with biochar is through soil compaction and soil fertility. The fields in the geographical area that he covers are characterized by a high degree of sandy soils, "so it is interesting that biochar can almost turn sandy soils into clay soil – if we get it distributed correctly." Kai further highlights the carbon removal potential and the likely reduced irrigation need, as well as the improved soil binding and nutrient supply, which altogether makes biochar "incredibly interesting to explore." While seeing several benefits from biochar use, Kai highlights climate and economy as two main reasons why farmers would engage with biochar:

> One wants to do something good for the climate – no one wants their children to be knee-deep in water. But if we don't have the economic sustainability included in this (…) if we don't have a healthy economic balance, then nothing else matters. (…) It is enormously important.

Thus, rather than being cast as climate culprits solely by external actors, Kai articulates that farmers themselves are concerned about the climate and feel compelled to attend to it (cf. Girvan 2018; Lovell, Bulkeley and Liverman 2009). However, he also insists that engaging with biochar has to be economically viable, and here the ability to account for carbon is increasingly important. In addition to the anticipated CO_2 tax, Kai states that the agrifood industry and financial institutions increasingly require farmers to engage in climate mitigation: "We just had our Climate Day last week," he says, referring to an event organized country-wide by an agricultural research institution in collaboration with local agricultural consultancies. He elaborates:

> The climate bottom line is also going to count in the bank, they say. So you have to have your finances in order, and you also need to have your climate [accounting] in order, and your responsibility [referring to ESG[12] principles]. Your [climate] footprint plays a larger and larger role.

Hence, farmers are increasingly expected to declare their emissions and consider initiatives to address them (cf. Lovell, Bulkeley and Liverman 2009; Lippert 2015). Like in Girvan's (2018) example of individual carbon footprints, the demand for farmer climate action does not come from a single central actor – the state – but from a diffuse set of actors within and beyond the sector. Thus, Kai (perhaps inadvertently) points to the neoliberal management regimes that farmers may become compelled to engage and comply with in order to benefit as biochar carbon removal providers. However, in assessing the pace of the roll-out of biochar in Denmark, he also raises concerns regarding the distribution of carbon removal benefits:

> It [the roll-out] depends on how quickly you get these commercial [pyrolysis] plants up and running, and what the economy of operations is, how steady they run, how they get the input, and what type of input they can handle. And then: what do we get for it? What will it cost? And the main factor in it: who will get the credit for the climate effect? Because (…) it becomes an economy in itself. That is the bottom line we are looking into.

Thus, farmers are positioned – and position themselves – as carbon subjects in a double sense: as corporate or individual actors who have to manage their carbon emissions and as potential providers of carbon removal who can potentially take part in the carbon economy. Yet, despite these overall ideas about how farmers could engage with biochar, how all this might unfold in practice is still not explicitly addressed.

Technology first

Continuing our discussion above, I asked Morten how he thinks biochar will be practically implemented. He hesitates: "Well, it is a discussion that we will have to have… (…) Now, we have to ensure that the technology can run on a large scale." Morten's hesitation indicates a general tendency in the sector to focus on "getting the technology right." This reflects a larger trend in Danish climate policy, where a preference for "technological solutions" prevails. Indicatively, Morten uses the term "pyrolysis" rather than "biochar," pointing to a tendency to focus on emerging pyrolysis start-ups rather than the "end users" of biochar. The focus on technology is also seen in the Danish Climate Programme, where this climate intervention is not labelled as biochar but as "brown biorefining (pyrolysis)" (DMCEU 2022, 161). Similarly, the national funding scheme supporting the development of the intervention is designated the Pyrolysis Grant,[13] while the Expert Group for a Green Tax Reform has suggested a subsidy scheme to further promote pyrolysis technology (EGGTR 2024, 10). Likewise, in network events for the emerging biochar sector, farmers are rarely present and only a fraction of the content is of relevance to them as end users (Hougaard 2024). Hence, although biochar is imagined to account for 25–33% of the reduction goal in the agricultural sector, there is little concern from central organizations for how the practice-technology will be practically implemented. Such framing of biochar as a technoscientific solution has been criticized in the social science literature. Scholars have argued that biochar is abstracted from its embeddedness in broader social and political agricultural practices, reinvented as a climate solution that can seamlessly be implemented across the world and applied to passive agricultural soils by an unidentified actor (Leach, Fairhead and Fraser 2012; Bezerra et al. 2019; Kon Kam King et al. 2018). Thus, while the focus in Danish policy and public debate is on "getting the technology right," the question of whether farmers are interested in adopting biochar and how they might engage with it is postponed. This creates discrepancies in the configuration of the biochar user.

At one end, Morten, Kai and other actors in the agricultural sector focus on the economic and climate mitigation reasons for why farmers might apply biochar. That is, they configure the farmer as an *output* (biochar) user. At the other end, farmers themselves are more concerned with the other side of production, namely providing *input* (biomass) material for the pyrolysis process. For instance, when asked about his interest in engaging with biochar, a young farmer only considered providing input in his answer: "Today I sell all of my hay to the biogas plant. If I should deliver to the pyrolysis plant in the future, they'll have to offer a really good price." Likewise, during a webinar on biochar organized by an agricultural research institution, questions in the chat revolved around opportunities for providing input material: "What will you pay per kg straw?", "Can biochar be made from sewage sludge?" and "What do I get in return after providing the straw?" Hence, for farmers who sell hay and straw by-products to biogas plants or for straw-based central district heating, pyrolysis plants are potential alternative buyers. Similarly, farms with large livestock productions generate large quantities of manure that is costly to get rid of. Each farmer needs a certain amount of land area to spread their manure (to comply with the so-called harmony rule[14] (Sommer and Knudsen 2021)), and many are concerned about reaching the limit of how much phosphorous they can apply to their land.[15] Being able to transform their manure into biochar, which is a much lighter product to transport than wet manure, provides the farmers with a welcome opportunity for manure treatment and disposal – and by selling it to others, it might even become a source of income instead of an expense. Thus, while some sectoral actors refer to *both* the pyrolysis process *and* biochar application when talking about biochar/pyrolysis, many farmers only hear "pyrolysis" and consider how they could provide input material to gain an extra income.

This tendency may be attributed to the excessive focus on pyrolysis and on "getting the technology right" in policy and public debate. However, technology is never just technology. It is entangled in and held up by various social, political, economic and ecological relations that determine the way in which – and whether – technology works. In their work on telecare technology for elderly people living at home, Sánchez-Criado et al. (2014) argue that technologies are not just "plug-n-play," but require installation for the specific user. They conceptualize "installations" as "those backstage, sociomaterial practices that usually only become visible during implementation and breakdown" (Sánchez-Criado et al. 2014, 697). While designers might have an imaginary user in mind when devising a technology (see Suchman 2002), Sánchez-Criado et al. (2014) draw on literature that looks at the co-construction of users and technologies to argue that neither services nor users pre-exist installation. They highlight that installations require careful work as well as technical, relational and contractual configurations for the user and the service to come into being as functional technology.

Likewise, I argue that biochar is not a "plug-n-play" technology. It cannot be readily applied to willing soils as a simple addition; it has to be produced, appropriated and carefully integrated into everyday farming practices. Using Sánchez-Criado et al.'s terminology, for biochar to work as a climate intervention, it will require careful work of installation in specific locations. First, the social, political, economic and ecological arrangements around the provision and handling of pyrolysis input material – be it straw, manure, wood chips, sewage or other biogenic material – must be taken into consideration and cannot be left until "after the technology works." Second, if the co-construction of users and technologies is to be taken seriously (Oudshoorn and Pinch 2003), the concrete spreading of biochar on soils must be considered upfront. This means integrating new agricultural practices for applying biochar, be it through a lime spreader, fertilizer spreader, chicken manure spreader or a pneumatic fertilizer spreader – as were the methods tested by Michael and Karen, whom we met in the opening of this chapter – or by digging it one metre below ground, as some researchers propose. Finally, with emerging practices of carbon accounting, for biochar "to work" towards national climate goals requires technical, relational and contractual work (Sánchez-Criado et al. 2014) to negotiate the climate benefits and potential carbon credits. This becomes clear when consulting people who have tested biochar on their farm.

Competing for the climate credit

Back on the farm, I follow Michael through the garage and into the garden, where we sit down on the terrace. Karen offers me something to drink as she joins us, while their four-year-old son roams around and wants her to read him a children's book. Michael reflects on the experiment in the adjacent field. "They couldn't really measure any difference last year," he states dryly and continues: "Perhaps if we don't get any rain the next three weeks, then maybe we might see some stripes where the biochar has been applied. (…) But up until now, we have not been able to see anything." Although the experiment was not to detect changes in plant production but to test application methods, the farmers would have liked to detect a crop improvement. Michael and Karen state that none of the application methods tested worked particularly well. "But it was fun to be part of," Karen adds. Karen and Michael echo Kai's description of farmers being concerned about the climate and wanting to act on this concern: "We actually wanted to contribute [to the climate agenda]," Karen says. "If we as farmers could contribute to this [challenge] of, how to create a climate-neutral agriculture…" Although she admits that other farmers might be more hesitant to engage with biochar due to economic concerns:

> It is very very flimsy [*flyvsk*], and farmers are perhaps in general not very good at dealing with stuff that cannot be readily monetarized. And before you can do that, you won't get farmers to join on a large scale.

Michael agrees and stresses the economic imperative for engaging with biochar:

> I think that the agricultural sector has to move beyond this idea that we do these sorts of things just to be nice. I will not use biochar on all of my fields just because we have a government who suddenly wants to wash their hands and say: 'See how good we are in Denmark, we are so CO_2 neutral.' It is the agricultural sector that does all the work, but if we still have to be scolded (…) then they can forget about it. At least there must be something in it for me, or some goodwill the other way (…) Otherwise, I doubt that [biochar] will revolutionize my yields to such a degree that I will employ it on a large scale.

Like Morten, Michael speaks to the wider societal discourse where farmers are cast as "climate culprits" (Kramer 2022); guilty carbon subjects who have to manage their carbon emissions. While biochar in this context offers a welcome technical solution that enables a continuation of current production (Kon Kam King et al. 2018), it also gives farmers the experience of being positioned as responsible for the burden of climate action. The farmers' interest in taking on such a burden depends, according to Michael and Karen, on whether they will receive some goodwill – recognition as "climate heroes" and economic benefits – thus making the topic of carbon accounting central. "If we in some way could be credited for applying this [biochar]," Karen says, "and be credited for storing CO_2, then I think (…) that it could become a big market." Hence, Karen does not just imagine that biochar could compensate for emissions elsewhere on the farm and thereby improve the farm's climate account. She also envisions that farmers could become carbon removal providers and sell carbon offsets on the voluntary carbon market, thereby becoming both climate heroes and gaining economically.

Karen's idea of engaging with the voluntary carbon market points to the concern that Kai raises, namely who should be credited for carbon removal and attain the economic benefits: the farmer providing the manure or biomass input, the pyrolysis plant or the farmer receiving the output material in the form of biochar? Different ideas for potential models circulate in the sector: farmers paying for receiving biochar, farmers receiving the same amount of biochar as they deliver input material for, farmers being ascribed the carbon stored in biochar while pyrolysis plants get carbon credits for replacing fossil fuels, farmers owning pyrolysis plants collectively and so on. These models are not necessarily compatible and raise concerns regarding additionality, temporality and leakage (Dalsgaard 2013; Lovell, Bulkeley and Liverman 2009; Gutiérrez 2011). Further, they anticipate that to become carbon farmers and carbon removal providers, farmers must adhere to neoliberal management regimes of measurements, standards and certification schemes (Asiyanbi, Ogar and Akintoye 2019; Paladino and Fiske 2017; Fletcher 2010).

The concrete distribution of the economic and climate benefits depends on the power relations between the different actors in the sector (Bryant, Dabhi and Böhm 2015; Ervine 2013; Lohmann 2008, 2009; Paladino and Fiske 2017). As a precursor to this negotiation, a recent conflict between biogas companies is illustrative – particularly as some pyrolysis entrepreneurs are experimenting with combining pyrolysis and biogas production in an integrated system: currently, some livestock farmers treat their manure through biogas plants and receive a part of the degasified manure in return for free in a form of exchange economy without financial transactions. The biogas plant then benefits from the sale of energy and from obtaining carbon credits for replacing fossil fuels. Thus, the increasing prices on energy and carbon credits have benefitted the biogas companies, while farmers who have entered long-term contracts are left in a less favourable situation. In lieu of the anticipated agricultural CO_2 tax, farmers are even more interested in obtaining the climate benefits and subtracting these from their farm emissions. While specific conflicts between farmers and biogas plants have been resolved, the memory of "being tricked" spills into the emerging constellations around farmer-pyrolysis collaboration and creates a sense of distrust and tension.

To add another layer of complexity to the emerging carbon accounting process, some pyrolysis entrepreneurs imagine producing synthetic fuels from the resulting bio-oil and syngas from the pyrolysis process for use in "fossil free aviation," thereby including the transport sector in the range of actors interested in claiming climate benefits. As Morten says, "I think it will be a crude political deal [*studehandel*] as to how that climate credit will be sold. (...) If you ask the pyrolysis people, they would say that the climate credit is theirs." At the time of writing, the first carbon credit from Danish biochar had been sold by a pyrolysis company on the voluntary carbon market to a non-domestic aviation company. Thus, credits are not only claimed in another sector but also in another country. Such a situation, where various actors claim the same carbon removal in their carbon accounting, reflects a long-standing practice in carbon markets and has been the subject of much critique in the literature (Ervine 2013; Gillenwater et al. 2007; Gutiérrez 2011; Knox-Hayes 2013; Lippert 2016; Lohmann 2009; MacKenzie 2009). Nevertheless, a Danish agricultural research institution has recently supported this practice, proposing that farmers be able to sell carbon offsets on the voluntary carbon market and that the removed carbon should also count towards Danish national climate goals. This leads to double claiming, where climate mitigation efforts are reported more than once (cf. Schneider and La Hoz Theuer 2019; Kreibich 2024). As mitigation activities overall thus seem larger than they are, such a configuration of biochar carbon farmers risks resulting in diminished emission reduction efforts.

Conclusion: making things stay the same

The emergence of carbon removal and net-zero goals invites us to revisit the concept of carbon accounting. As climate action is increasingly envisioned to take place within the Global North, new actors are enrolled as double carbon subjects: both as climate culprits that must manage their emissions and as potential carbon removal providers that must adhere to neoliberal management regimes. While the Global North's domestication of carbon removal activities may in theory avert problematic neo-colonial climate relations (Navarro 2022; Zografos and Robbins 2020; Bachram 2004), biochar is still positioned as a technoscientific or techno-political practice (Leach, Fairhead and Fraser 2012; Bezerra et al. 2019; see also Lippert 2016), resulting in a disconnect between farmers' and other actors' expectations about their role in biochar engagements. As markets, labour and power relations are being reconfigured, ideas about how economic and climate benefits should be distributed are not aligned, and issues of additionality, temporality and leakage seem to remain (Lovell, Bulkeley and Liverman 2009; Gutiérrez 2011).

Discrepancies over how the biochar farmer should be configured are illustrative of user configurations as iterative and ongoing processes (Oudshoorn and Pinch 2003; Sánchez-Criado et al. 2014). Yet, discrepancies may result in double claiming, giving the impression that climate mitigation efforts are larger than they are. This is not to argue that if carbon farmers were "properly configured" and accounting was "done right" there would be no problem – a large body of literature testifies to the contrary (Dalsgaard 2013; Ervine 2012; Gutiérrez 2011; Knox-Hayes 2013; Lohmann 2008, 2009). Rather, I argue that when relying on carbon accounting and market-based climate policy, new forms of carbon subjects emerge and seek ways to tap into and benefit from the carbon market. The introduction of biochar in Danish climate policy illustrates that "counting carbon" encourages actors in and around the agricultural sector to establish new complex technical systems in order to keep the overall agricultural model and their production capacity in place. Instead of prioritizing substantial reductions, for instance, through downscaling livestock production (Brichet, Brieghel and Hastrup 2023), sectoral actors endorse the government's reliance on untested carbon removal – even though this requires massive investments in pyrolysis plants and even though the benefits for farmers are uncertain. Biochar thus emerges as a "technical adjustment" (Günel 2019) that promises to fix the problem of climate change without changing the social and political status quo. Hence, in line with Schyberg (this volume), this case illustrates that carbon as a concept and a material form makes people and institutions take on new roles and practices in order to make other things stay the same.

Acknowledgements

Thanks to my interlocutors in the Danish biochar sector for taking the time for visits, interviews and follow-up conversations. Thanks to participants at the Sociocultural Carbon (SOCCAR) Symposium hosted at IT University of Copenhagen, and the Nature, Environment and Climate (NEC) research group at the Department of Anthropology, University of Copenhagen, for useful feedback on an earlier draft of this chapter. Finally, I wish to thank the book editors – Steffen Dalsgaard, Katinka Schyberg, Andy Lautrup and Ingmar Lippert – for competent guidance and constructive comments. The research benefitted from funding from the Swedish Research Council FORMAS (Grant number 2019-01953).

Notes

1 All names of people are pseudonymized, while organizations are anonymised.
2 In Denmark, the terms "biochar" and "pyrolysis" are used interchangeably, though they refer to different processes. In this chapter, I mainly use "biochar," since this is the predominant term in academic literature on carbon removal (e.g. Minx et al. 2018).
3 While CDR methods are often considered "technologies," I use the term "practice-technologies" to highlight that the capturing and storing of carbon rely on everyday work practices as well as social, economic and ecological relations.

4 CO_2e (or CO_2eq) is the internationally used abbreviation for greenhouse gas emissions, covering various gasses converted into their equivalent CO_2-emissions by estimating their global warming potential over a period of 100 years (see Lippert 2012; MacKenzie 2009).
5 It is unclear how large a share of the anticipated carbon capture and storage (CCS) is expected to be based on biogenic CO_2 (and thus constitute carbon removal) and how much on fossil emissions (and thus constitute "mere" reductions).
6 In later climate programmes, the share of anticipated mitigation through CCS has been adjusted, and while the amount of expected removals through biochar until recently has remained the same, it was in the 2024 Climate Programme downscaled to 0.3 Mt CO_2e (DMCEU 2024).
7 While it is most often considered relevant for the agricultural sector due to its soil-improving qualities, biochar can also be used in construction work and in urban green areas.
8 Many Danish farmers refer to their farm as a "business," thus indicating what I see as an increasing "corporatization" of the farm.
9 While *double counting* means that emission reductions or removals are counted twice in, for instance, national emission accounts, *double claiming* occurs when emission reductions or removals are counted both in national (territorial) accounts and in corporate reporting through the use of carbon credits for voluntary offsetting (Schneider and La Hoz Theuer 2019).
10 Fairhead, Leach and Kojo (2012) use the term *Anthropogenic Dark Earths* to highlight that the practice was also used in other parts of the world.
11 Water-retainment has, however, been documented on coarse sandy soils (Bruun et al. 2022; 2014; Elsgaard et al. 2022).
12 ESG is an acronym for "Environmental, Social and Governance" and is used to evaluate corporate sustainability performance. In recent years, it has entered the agricultural sector as a parameter that financial institutions use to assess farm investments and loans.
13 The funding body does, however, require that grant applicants engage with the entire supply chain.
14 The "harmony rule" determines how much manure the farmers can apply to their fields each year, with the aim of reducing the risk of nutrient surplus and run-off that can damage the aquatic environment. The rules determine a minimum land area per livestock unit, which the farmer can either own or have access to otherwise (DAC 2024; Sommer and Knudsen 2021).
15 While phosphorous is considered a scarce resource on a global scale, its accumulation in agricultural soils due to high levels of fertilizer (manure) application in Denmark implies a risk of phosphorous leakage and thus is considered an environmental hazard (Jarvie et al. 2015).

References

Armstrong, Chris, and Duncan McLaren. 2022. "Which Net Zero? Climate Justice and Net Zero Emissions". *Ethics & International Affairs* 36 (4): 505–26. https://doi.org/10.1017/S0892679422000521

Asiyanbi, Adeniyi P., Edwin Ogar, and Oluyemi A. Akintoye. 2019. "Complexities and Surprises in Local Resistance to Neoliberal Conservation: Multiple Environmentalities, Technologies of the Self and the Poststructural Geography of Local Engagement with REDD+". *Political Geography* 69: 128–38. https://doi.org/10.1016/j.polgeo.2018.12.008

Azzi, Elias S., Erik Karltun, and Cecilia Sundberg. 2021. "Small-Scale Biochar Production on Swedish Farms: A Model for Estimating Potential, Variability, and

Environmental Performance". *Journal of Cleaner Production* 280 (2): 124873. https://doi.org/10.1016/j.jclepro.2020.124873

Bachram, Heidi. 2004. "Climate Fraud and Carbon Colonialism: The New Trade in Greenhouse Gases". *Capitalism Nature Socialism* 15 (4): 5–20. https://doi.org/10.1080/1045575042000287299

Battersby, Francesca, Richard J. Heap, Adam C. Gray, Mark Workman, and Finn Strivens. 2022. "The Role of Corporates in Governing Carbon Dioxide Removal: Outlining a Research Agenda". *Frontiers in Climate* 4: 686762. https://doi.org/10.3389/fclim.2022.686762

Bezerra, Joana. 2015. "Terra Preta de Índio and Amazonian History". In *The Brazilian Amazon: Politics, Science and International Relations in the History of the Forest*, 15–58. Cham: Springer. https://doi.org/10.1007/978-3-319-23030-6

Bezerra, Joana, Esther Turnhout, Isabel Melo Vasquez, Tatiana Francischinelli Rittl, Bas Arts, and Thomas W. Kuyper. 2019. "The Promises of the Amazonian Soil: Shifts in Discourses of Terra Preta and Biochar". *Journal of Environmental Policy and Planning* 21 (5): 623–35. https://doi.org/10.1080/1523908X.2016.1269644

Bis, Zbigniew, Rafał Kobyłecki, Mariola Ścisłowska, and Robert Zarzycki. 2018. "Biochar – Potential Tool to Combat Climate Change and Drought". *Ecohydrology and Hydrobiology* 18 (4): 441–53. https://doi.org/10.1016/j.ecohyd.2018.11.005

Brichet, Nathalia, Signe Brieghel, and Frida Hastrup. 2023. "Feral Kinetics and Cattle Research Within Planetary Boundaries". *Animals* 13 (5): 802. https://doi.org/10.3390/ani13050802

Bruckermann, Charlotte. 2022. ""There's an App for That!": Ordering Claims on Natural Resources through Individual Carbon Accounts in China". *Capitalism Nature Socialism* 33 (4): 95–114. https://doi.org/10.1080/10455752.2022.2089705

Bruun, Esben W., Dorette Müller-Stöver, Betina Nørgaard Pedersen, Line Vinther Hansen, and Carsten Tilbæk Petersen. 2022. "Ash and Biochar Amendment of Coarse Sandy Soil for Growing Crops under Drought Conditions". *Soil Use and Management* 38 (2): 1280–92. https://doi.org/10.1111/sum.12783

Bruun, Esben W., C.T. Petersen, E. Hansen, J.K. Holm, and H. Hauggaard-Nielsen. 2014. "Biochar Amendment to Coarse Sandy Subsoil Improves Root Growth and Increases Water Retention". *Soil Use and Management* 30 (1): 109–18. https://doi.org/10.1111/sum.12102

Bryant, Gareth, Siddhartha Dabhi, and Steffen Böhm. 2015. ""Fixing" the Climate Crisis: Capital, States, and Carbon Offsetting in India". *Environment and Planning A* 47 (10): 2047–63. https://doi.org/10.1068/a130213p

Buylova, Alexandra, Mathias Fridahl, Naghmeh Nasiritousi, and Gunilla Reischl. 2021. "Cancel (Out) Emissions? The Envisaged Role of Carbon Dioxide Removal Technologies in Long-Term National Climate Strategies". *Frontiers in Climate* 3 (675499). https://doi.org/10.3389/fclim.2021.675499

Carton, Wim, Adeniyi Asiyanbi, Silke Beck, Holly Jean Buck, and Jens Friis Lund. 2020. "Negative Emissions and the Long History of Carbon Removal". *Wiley Interdisciplinary Reviews: Climate Change* 11 (6): e671. https://doi.org/10.1002/wcc.671

Carton, Wim, Inge-Merete Hougaard, Nils Markusson, and Jens Friis Lund. 2023. "Is Carbon Removal Delaying Emission Reductions?" *WIREs Climate Change* 14 (4): e826. https://doi.org/10.1002/wcc.826

Carton, Wim, Jens Friis Lund, and Kate Dooley. 2021. "Undoing Equivalence: Rethinking Carbon Accounting for Just Carbon Removal". *Frontiers in Climate* 3: 664130. https://doi.org/10.3389/fclim.2021.664130

Creutzig, Felix, Karl-Heinz Erb, Helmut Haberl, Christian Hof, Carol Hunsberger, and Stephanie Roe. 2021. "Considering Sustainability Thresholds for BECCS in IPCC and Biodiversity Assessments". *GCB Bioenergy* 13 (4): 510–15. https://doi.org/10.1111/gcbb.12798

DAC. 2024. "Harmoniregler [Harmony rules]". *Danish Agricultural Agency (DAC)*. https://lbst.dk/landbrug/goedning/husdyrgoedning-og-anden-organisk-goedning/harmoniregler

DAFC. 2022. *Climate-Neutral 2050*. Bruxelles: Danish Agriculture and Food Council (DAFC). https://agricultureandfood.dk/media/uofnbtfb/climate-neutral-2050-dafc.pdf

Dalsgaard, Steffen. 2013. "The Commensurability of Carbon: Making Value and Money of Climate Change". *HAU: Journal of Ethnographic Theory* 3 (1): 80–98. https://doi.org/10.14318/hau3.1.006

Danish Government. 2019. *Aftale Om Klimalov Af 6. December 2019 [Agreement on Climate Act of December 6, 2019]*. Copenhagen: Danish Ministry for Climate, Energy and Utilities (DMCEU).

Danish Government. 2021. *Aftale Om Grøn Omstilling Af Dansk Landbrug [Agreement on the Green Transition of Danish Agriculture]*. Copenhagen: Ministry of Finance.

Danish Government. (2022). Regeringsgrundlag: Ansvar for Danmark [Foundation for Government: Responsibility for Denmark]. Prime Minister's Office. https://www.stm.dk/statsministeriet/publikationer/regeringsgrundlag-2022/

Danish Government, DAFC, Danish Society for Nature Conservation, Trade Union NNF, Danish Metal Workers Union, Danish Industry, and KL - Local Government Denmark. 2024. *Aftale om et Grønt Danmark [Agreement on a Green Denmark]*. Danish Government. https://www.regeringen.dk/media/13261/aftale-om-et-groent-danmark.pdf

DEA. 2023. *Klimastatus og -fremskrivning 2023 [Climate status and projections 2023]*. Copenhagen: Danish Energy Agency (DEA).

DMCEU. 2020. *Klimaprogram 2020 [Climate Programme 2020]*. Copenhagen: Danish Ministry for Climate, Energy and Utilities (DMCEU).

DMCEU. 2022. *Klimaprogram 2022 [Climate Programme 2022]*. Copenhagen: Danish Ministry for Climate, Energy and Utilities (DMCEU).

DMCEU. 2024. "The Danish Government and Parties in "Green Tripartite" Agree on a Ground-Breaking Model for a Carbon Tax on Food and Agriculture Production". *Danish Ministry of Climate, Energy and Utilities (DMCEU)*. https://www.en.kefm.dk/news/news-archive/2024/jul/the-danish-government-and-parties-in-"green-tripartite"-agree-on-a-ground-breaking-model-for-a-carbon-tax-on-food-and-agriculture-production

Dumbrell, Nikki P., Marit E. Kragt, and Fiona L. Gibson. 2016. "What Carbon Farming Activities Are Farmers Likely to Adopt? A Best–Worst Scaling Survey". *Land Use Policy* 54: 29–37. https://doi.org/10.1016/j.landusepol.2016.02.002

Dyrhauge, Helene. 2021. "Political Myths in Climate Leadership: The Case of Danish Climate and Energy Pioneership". *Scandinavian Political Studies* 44 (1): 13–33. https://doi.org/10.1111/1467-9477.12185

EGGTR. 2024. "Green Tax Reform: Final Report." Expert Group for a Green Tax Reform (EGGTR), Danish Ministry of Taxation (DMT). https://skm.dk/media/tngh1b4r/green-tax-reform-final-report.pdf

Elsgaard, Lars, Anders Peter S. Adamsen, Henrik B. Møller, Anne Winding, Uffe Jørgensen, Esben Ø. Mortensen, Emmanuel Arthur, et al. 2022. "Knowledge Synthesis on Biochar in Danish Agriculture". 208. DCA Report. Aarhus University.

Ervine, Kate. 2012. "The Politics and Practice of Carbon Offsetting: Silencing Dissent". *New Political Science* 34 (1): 1–20. https://doi.org/10.1080/07393148.2012.646017

Ervine, Kate. 2013. "Carbon Markets, Debt and Uneven Development". *Third World Quarterly* 34 (4): 653–70. https://doi.org/10.1080/01436597.2013.786288

European Commission. 2021. "Carbon Farming". *European Commission*. https://climate.ec.europa.eu/eu-action/sustainable-carbon-cycles/carbon-farming_en

European Commission. 2023. "CAP at a Glance". *European Commission*. https://agriculture.ec.europa.eu/common-agricultural-policy/cap-overview/cap-glance_en

Fairhead, James, Melissa Leach, and Amanor Kojo. 2012. "Anthropogenic Dark Earths and Africa. A Political Agronomy of Research Disjunctures". In *Contested Agronomy: Agricultural Research in a Changing World*, edited by James Sumberg and John Thompson, 64–77. Taylor & Francis Group.

Fletcher, Robert. 2010. "Neoliberal Environmentality: Towards a Poststructuralist Political Ecology of the Conservation Debate". *Conservation and Society* 8 (3): 171. https://doi.org/10.4103/0972-4923.73806

Fuss, Sabine, William F. Lamb, Max W. Callaghan, Jérôme Hilaire, Felix Creutzig, Thorben Amann, Tim Beringer, et al. 2018. "Negative Emissions - Part 2: Costs, Potentials and Side Effects". *Environmental Research Letters* 13 (6). https://doi.org/10.1088/1748-9326/aabf9f

Gillenwater, Michael, Derik Broekhoff, Mark Trexler, Jasmine Hyman, and Rob Fowler. 2007. "Policing the Voluntary Carbon Market". *Nature Climate Change* 1 (711): 85–87. https://doi.org/10.1038/climate.2007.58

Girvan, Anita. 2018. *Carbon Footprints as Cultural-Ecological Metaphors*. Routledge Environmental Humanities. London New York: Routledge, Taylor & Francis Group.

Günel, Gökce. 2019. *Spaceship in the Desert: Energy, Climate Change and Urban Design in Abu Dhabi*. Durham: Duke University Press.

Gutiérrez, María. 2011. "Making Markets out of Thin Air: A Case of Capital Involution". *Antipode* 43 (3): 639–61. https://doi.org/10.1111/j.1467-8330.2011.00884.x

Gutiérrez, María. 2016. "Forest Carbon Sinks Prior to REDD: A Brief History of Their Role in the Clean Development Mechanism". In The Carbon Fix: Forest Carbon, Social Justice, and Environmental Governance, edited by Stephanie Paladino and Shirley J. Fiske, 84–97. Routledge.

Hagenbo, Andreas, Clara Antón-Fernández, Ryan M. Bright, Daniel Rasse, and Rasmus Astrup. 2022. "Climate Change Mitigation Potential of Biochar from Forestry Residues under Boreal Condition". *Science of The Total Environment* 807 (3): 151044. https://doi.org/10.1016/j.scitotenv.2021.151044

Hansson, Anders, Simon Haikola, Mathias Fridahl, Pius Yanda, Edmund Mabhuye, and Noah Pauline. 2021. "Biochar as Multi-Purpose Sustainable Technology: Experiences from Projects in Tanzania". *Environment, Development and Sustainability* 23: 5182–5214. https://doi.org/10.1007/s10668-020-00809-8

Hastrup, Frida, Nathalia Brichet, and Liza Rosenbaum Nielsen. 2022. "Sustainable Animal Production in Denmark: Anthropological Interventions". *Sustainability (Switzerland)* 14 (9): 5584. https://doi.org/10.3390/su14095584

Hougaard, Inge-Merete. 2024. "Enacting Biochar as a Climate Solution in Denmark". *Environmental Science & Policy* 152: 103651. https://doi.org/10.1016/j.envsci.2023.103651

IPCC. 2018. *Global Warming of 1.5°C*. Intergovernmental Panel on Climate Change (IPCC).

IPCC. 2022. *Climate Change 2022. Mitigation of Climate Change. Summary for Policymakers*. Intergovernmental Panel on Climate Change (IPCC).

Jänicke, Martin, and Rüdiger K.W. Wurzel. 2019. "Leadership and Lesson-Drawing in the European Union's Multilevel Climate Governance System". *Environmental Politics* 28 (1): 22–42. https://doi.org/10.1080/09644016.2019.1522019

Jarvie, Helen P., Andrew N. Sharpley, Don Flaten, Peter J.A. Kleinman, Alan Jenkins, and Tarra Simmons. 2015. "The Pivotal Role of Phosphorus in a Resilient Water-Energy-Food Security Nexus". *Journal of Environmental Quality* 44 (4): 1049–62. https://doi.org/10.2134/jeq2015.01.0030

Juul Christiansen, Flemming. 2020. "The Liberal Party: From Agrarian and Liberal to Centre-Right Catch-All". In *The Oxford Handbook of Danish Politics*, edited by Peter Munk Christiansen, Jørgen Elklit, and Peter Nedergaard, 295–312. Oxford University Press. https://doi.org/10.1093/oxfordhb/9780198833598.013.23

Knox-Hayes, Janelle. 2013. "The Spatial and Temporal Dynamics of Value in Financialization: Analysis of the Infrastructure of Carbon Markets". *Geoforum* 50: 117–28. https://doi.org/10.1016/j.geoforum.2013.08.012

Kon Kam King, Juliette, Céline Granjou, Juliette Fournil, and Lauric Cecillon. 2018. "Soil Sciences and the French 4 per 1000 Initiative - The Promises of Underground Carbon". *Energy Research and Social Science* 45: 144–52. https://doi.org/10.1016/j.erss.2018.06.024

Kramer, Cæcilie Kildahl. 2022. "Kamp Om Jorden: Et Antropologisk Blik På "jordens Folk" i Midten Af Klima- Og Miljøkampen [The Struggle for Land: An Anthropological View on "People of the Earth" in the Middle of the Climate and Environmental Struggle]". *Jordens Folk* 57 (1): 36–50.

Kreibich, Nicolas. 2024. "Toward Global Net Zero: The Voluntary Carbon Market on Its Quest to Find Its Place in the post-Paris Climate Regime". *WIREs Climate Change*, May, e892. https://doi.org/10.1002/wcc.892

Leach, Melissa, James Fairhead, and James Fraser. 2012. "Green Grabs and Biochar: Revaluing African Soils and Farming in the New Carbon Economy". *Journal of Peasant Studies* 39 (2): 285–307. https://doi.org/10.1080/03066150.2012.658042

Lippert, Ingmar. 2012. "Carbon Dioxide". In *Encyclopedia of Consumption and Waste: The Social Science of Garbage*, edited by Carl Zimring and William Rathje, 105–7. Thousand Oaks: SAGE Publications, Inc. https://doi.org/10.4135/9781452218526

Lippert, Ingmar. 2015. "Environment as Datascape: Enacting Emission Realities in Corporate Carbon Accounting". *Geoforum* 66: 126–35. https://doi.org/10.1016/j.geoforum.2014.09.009

Lippert, Ingmar. 2016. "Corporate Carbon Footprinting as Techno-Political Practice". In *The Carbon Fix: Forest Carbon, Social Justice, and Environmental Governance*, edited by Stephanie Paladino and Shirley J. Fiske, 131–42. Routledge.

Lohmann, Larry. 2008. "Carbon Trading, Climate Justice and the Production of Ignorance: Ten Examples". *Development* 51 (3): 359–65. https://doi.org/10.1057/dev.2008.27

Lohmann, Larry. 2009. "Regulation as Corruption in the Carbon Offset Markets". In *Upsetting the Offset: The Political Economy of Carbon Markets*, edited by Steffen Böhm and Siddhartha Dabhi, 175–91. London: MayFlyBook.

Lovell, Heather, Harriet Bulkeley, and Diana Liverman. 2009. "Carbon Offsetting: Sustaining Consumption?" *Environment and Planning A* 41 (10): 2357–79. https://doi.org/10.1068/a40345

Lovell, Heather, and Donald Mackenzie. 2011. "Accounting for Carbon: The Role of Accounting Professional Organisations in Governing Climate Change". *Antipode* 43 (3): 704–30. https://doi.org/10.1111/j.1467-8330.2011.00883.x

Lund, Jens Friis, Nils Markusson, Wim Carton, and Holly Jean Buck. 2023. "Net Zero and the Unexplored Politics of Residual Emissions". *Energy Research & Social Science* 98. https://doi.org/10.1016/j.erss.2023.103035

MacKenzie, Donald. 2009. "Making Things the Same: Gases, Emission Rights and the Politics of Carbon Markets". *Accounting, Organizations and Society* 34 (3–4): 440–55. https://doi.org/10.1016/j.aos.2008.02.004

Markusson, Nils, Duncan McLaren, and David Tyfield. 2018. "Towards a Cultural Political Economy of Mitigation Deterrence by Negative Emissions Technologies (NETs)". *Global Sustainability* 1: E10. https://doi.org/10.1017/sus.2018.10

McLaren, Duncan, David P. Tyfield, Rebecca Willis, Bronislaw Szerszynski, and Nils O. Markusson. 2019. "Beyond "Net-Zero": A Case for Separate Targets for Emissions Reduction and Negative Emissions". *Frontiers in Climate* 1: 4. https://doi.org/10.3389/fclim.2019.00004

Minx, Jan, William F. Lamb, Max W. Callaghan, Sabine Fuss, Jérôme Hilaire, Felix Creutzig, Thorben Amann, et al. 2018. "Negative Emissions - Part 1: Research

Landscape and Synthesis". *Environmental Research Letters* 13 (6): 063001. https://doi.org/10.1088/1748-9326/aabf9b

Nader, Laura. 1972. "Up the Anthropologist: Perspectives Gained from Studying Up". In *Reinventing Anthropology*, edited by D. Hymes, 284–311. New York: Pantheon.

Navarro, Rebecca. 2022. "Climate Finance and Neo-Colonialism: Exposing Hidden Dynamics". In *The Political Economy of Climate Finance: Lessons from International Development*, edited by Corrine Cash and Larry A. Swatuk. Cham: Palgrave Macmillan. https://doi.org/10.1007/978-3-031-12619-2_8

Okyere, Charles Yaw, and Lukas Kornher. 2023. "Carbon Farming Training and Welfare: Evidence from Northern Ghana". *Land Use Policy* 134: 106932. https://doi.org/10.1016/j.landusepol.2023.106932

Otte, Pia Piroschka, and Jostein Vik. 2017. "Biochar Systems: Developing a Socio-Technical System Framework for Biochar Production in Norway". *Technology in Society* 51: 34–45. https://doi.org/10.1016/j.techsoc.2017.07.004

Oudshoorn, Nelly, and Trevor Pinch. 2003. "Introduction: How Users and Non-Users Matter". In *How Users Matter: The Co-Construction of Users and Technologies*, edited by Nelly Oudshoorn and Trevor Pinch, 1–25. The MIT Press.

Paladino, Stephanie, and Shirley J. Fiske, eds. 2017. *The Carbon Fix: Forest Carbon, Social Justice, and Environmental Governance*. New York London: Routledge.

Paul, Carsten, Bartosz Bartkowski, Cenk Dönmez, Axel Don, Stefanie Mayer, Markus Steffens, Sebastian Weigl, Martin Wiesmeier, André Wolf, and Katharina Helming. 2023. "Carbon Farming: Are Soil Carbon Certificates a Suitable Tool for Climate Change Mitigation?" *Journal of Environmental Management* 330: 117142. https://doi.org/10.1016/j.jenvman.2022.117142

Pratt, Kimberley, and Dominic Moran. 2010. "Evaluating the Cost-Effectiveness of Global Biochar Mitigation Potential". *Biomass and Bioenergy* 34 (8): 1149–58. https://doi.org/10.1016/j.biombioe.2010.03.004

Sánchez-Criado, Tomás, Daniel López, Celia Roberts, and Miquel Domènech. 2014. "Installing Telecare, Installing Users: Felicity Conditions for the Instauration of Usership". *Science Technology and Human Values* 39 (5): 694–719. https://doi.org/10.1177/0162243913517011

Schenuit, Felix, Rebecca Colvin, Mathias Fridahl, Barry Mcmullin, Andy Reisinger, Daniel L. Sanchez, Stephen M. Smith, Asbjørn Torvanger, and Anita Wreford. 2021. "Carbon Dioxide Removal Policy in the Making: Assessing Developments in 9 OECD Cases". *Frontiers in Climate* 3: 638805. https://doi.org/10.3389/fclim.2021.638805

Schmidt, Hans Peter, Andrés Anca-Couce, Nikolas Hagemann, Constanze Werner, Dieter Gerten, Wolfgang Lucht, and Claudia Kammann. 2019. "Pyrogenic Carbon Capture and Storage". *Global Change Biology: Bioenergy* 11 (4): 573–91. https://doi.org/10.1111/gcbb.12553

Schneider, Lambert, and Stephanie La Hoz Theuer. 2019. "Environmental Integrity of International Carbon Market Mechanisms under the Paris Agreement". *Climate Policy* 19 (3): 386–400. https://doi.org/10.1080/14693062.2018.1521332

Smith, Pete. 2016. "Soil Carbon Sequestration and Biochar as Negative Emission Technologies". *Global Change Biology* 22 (3): 1315–24. https://doi.org/10.1111/gcb.13178

Soentgen, Jens, Klaus Hilbert, Carolin von Groote-Bidlingmaier, Gabriele Herzog-Schröder, Eije Erich Pabst, and Sabine Timpf. 2017. "Terra Preta de Índio: Commodification and Mythification of the Amazonian Dark Earths". *Gaia* 26 (2): 136–43. https://doi.org/10.14512/gaia.26.2.18

Sommer, Sven G., and Leif Knudsen. 2021. "Impact of Danish Livestock and Manure Management Regulations on Nitrogen Pollution, Crop Production, and Economy". *Frontiers in Sustainability* 2: 658231. https://doi.org/10.3389/frsus.2021.658231

Stanley, Theo. 2024. "Carbon "Known Not Grown": Reforesting Scotland, Advanced Measurement Technologies, and a New Frontier of Mitigation Deterrence". *Environmental Science & Policy* 151: 103636. https://doi.org/10.1016/j.envsci.2023.103636

Suchman, Lucy. 2002. "Located Accountabilities in Technology Production". *Scandinavian Journal of Information Systems* 14: 91–105.

Tammeorg, Priit, Päivi Soronen, Anu Riikonen, Esko Salo, Suvi Tikka, Minja Koivunen, Anna-Reetta Salonen, Topi Kopakkala, and Mikko Jalas. 2021. "Co-Designing Urban Carbon Sink Parks: Case Carbon Lane in Helsinki". *Frontiers in Environmental Science* 9: 672468. https://doi.org/10.3389/fenvs.2021.672468

Thomsen, Tobias Pape, Magnus Bo Karlsson, and Andreas Kamp. 2023. *Styrket Grundlag for Vurdering Af Klimaeffekter Ved Pyrolyse Af Tre Forskellige Typer Af Biomasse - et Arbejds - Og Baggrundsnotat [Strengthened Basis for Evaluating Climate Effects of Pyrolysis of Three Different Types of Biomass - a Work and Background Note]*. Roskilde University.

Woolgar, Steve. 1990. "Configuring the User: The Case of Usability Trials". *The Sociological Review* 38 (1_suppl): 58–99. https://doi.org/10.1111/j.1467-954x.1990.tb03349.x

Zografos, Christos, and Paul Robbins. 2020. "Green Sacrifice Zones, or Why a Green New Deal Cannot Ignore the Cost Shifts of Just Transitions". *One Earth* 3 (5): 543–46. https://doi.org/10.1016/j.oneear.2020.10.012

4 The footprint of anarchy
Counting carbon in the Church

Katinka A. Schyberg

Introduction

At the beginning of 2020, the Danish People's Church was encouraged by two sides to take action to help achieve the Danish government's official goal of reducing Denmark's carbon emissions by 70% by 2030. First, the ten bishops within the Church stated in the minutes from their triannual meeting that, in their opinion, the Danish Church, being a public institution, should support the government's reduction goal, as well as take measures to reduce the Church's own emissions. Then, a few weeks later, the political party Alternativet ("The Alternative") proposed a new law that would make it mandatory for the Church to lease its land exclusively to organic farming and reforestation projects. The Danish Church's responsibility for contributing to the reduction of national emissions was thus directly addressed both from within, by the bishops, and from outside, by politicians in Parliament. Considering that the Church is a public institution, it is perhaps not surprising that such actors find the goals set by the national government to apply to the Church. The Church is written into the Danish Constitution as one of the four pillars of Danish society, and although it is partly funded by membership taxes, it also receives substantial funds from the State for, among other things, the employment of clergy (i.e. priests, bishops and deans). Furthermore, other private organizations, public institutions in Denmark and abroad, and even national Churches in countries such as Norway, England and Sweden[1] have, in recent years, committed to specific carbon reduction goals and made more or less detailed action plans on how to reach them. However, both the statement made by the bishops and the law proposed by Alternativet spurred significant controversy – the suggested bill was defeated by a large majority of parties represented in Parliament and the bishops had to backpedal after being accused of misusing their bishoprics (Bramming 2020; Skov Hansen 2020). The statements spurred controversy for similar, though slightly different, reasons related to the organizational ethos of the Church. The controversies revolved around the question of whether anyone – be they bishops or the State – can or should speak, make decisions or take responsibility on behalf of the entire Danish Church. Most Church actors would reply in the negative: the Church is a decentralized organization, a "well-ordered anarchy," as it is popularly called by Church actors. No one should be able to dominate or speak on its behalf. However, in early 2022, something called Folkekirkens Grønne Omstilling ("The Green Transition of the Danish People's Church") nonetheless appeared in the news and on a website of the same name. The Church had formulated a collective response to climate change in which it promised to play its part in the national effort to green Danish society. If the controversies in January 2020 made it seem as though such a response was an institutional impossibility, something had changed.

This chapter sheds light on what happened as carbon was introduced as a resource to think with, evaluate by and act upon in a specific socio-cultural context – that of the Danish People's Church. The chapter specifically attends to how the prospect of calculating and displaying carbon footprints played a significant role in the negotiations of State-Church relations in Denmark in the years 2020–2022 when this Scandinavian country – including its national Church – tried to live up to the reduction goals outlined in its 2021 national climate law. In the chapter, I show what happened as the Church both resisted and was drawn into the logic of carbon footprinting, and I show how the prospect of such footprinting influenced the relationship between the State and the Church – a relationship that, for the most part, goes unnoticed but which became a matter of public scrutinization and contestation as the need for a green transition in Denmark also became a Church responsibility. The chapter shows how accounting for carbon emissions was used in the service of achieving what is otherwise deemed impossible: namely, to speak on behalf of the entire Church without compromising either its internal "religious freedom" (*trosfrihed*) or its ethos of "spaciousness" (*rummelighed*) – the values that underpin the organization of the Church. I argue that the method of calculating and displaying carbon footprints as a response to climate change offered the Church a way to simultaneously claim responsibility while not having to compromise on the principle of decentralization.

I end the chapter by speculating that the introduction of such a carbon logic may indeed come to change the Church more substantially than it was supposed to. As critical scholars of environmental accounting have argued, accounting practices cannot be considered a neutral device but are practices that affect the reality they account for (MacKenzie 2009; Lovell and MacKenzie 2011; Lippert 2015; Blok 2011). While such scholars are mostly concerned with how accounting for carbon emissions as a way of governing climate change shapes the reality of climate change and the politics of mitigating it, in this chapter, I am concerned with how carbon accounting might also shape the Church's organization. Because not only do these calculations influence how the Church approaches climate change, they might also change the organizational structure and, indeed, principles of the Church. As a first step, carbon footprinting may serve to preserve the much cherished, internal "ordered anarchy" (*velordnede anarki*) of the Church. But in the longer run, I hypothesize, it may also threaten it, because giving an account of oneself (Butler 2005) tends to not only have communicative effects but also affect the one doing the accounting. To account for itself by way of a carbon footprint might not only produce the illusion of a Church as a unity – an illusion that serves the purpose of keeping the State at bay and of protecting internal heterogeneity – but also move it in the direction of becoming so. As a form of accounting, carbon footprinting might participate in organizing the Church more uniformly than Church actors would in fact like it to. Hence, what I suggest is that climate change mitigation in the form of carbon footprinting might have an unintentional organizing effect on the Church so that it is not only the Church that organizes climate change mitigation but also climate change mitigation that organizes the Church. To account for anarchy might also be a way of organizing it.

In the first part of the chapter, I interrogate why it seemed so impossible for the Danish Church to commit to national carbon emission reduction goals in 2020. To understand how carbon numbers were mobilized in the Church in 2022, it is necessary to know what kind of Church we are talking about and what its relationship is to the State. In the coming pages, I unpack some aspects of the Church's organizational structure and the values underpinning it. Here, I show how this specific socio-cultural context for engaging with carbon footprinting proposed this method as not only a tool of governance but also of un-governance. I am specifically concerned with three key values that underpin Church organization and with which the controversies around carbon emission goals collided: "freedom" (*frihed*), "equality" (*lighed*) and "spaciousness" (*rummelighed*). In the second part of the chapter, I describe how the prospect of calculating and displaying carbon footprints came to play a significant role in the Church's efforts (and those of the State) to formulate a response to the climate crisis.

10 bishops = 10 greengrocers

In January 2020, the ten bishops of the Danish People's Church stated in the minutes from their tri-annual meeting that, being a public institution, the People's Church should support the Danish government's goal to reduce Denmark's carbon emissions by 70% by 2030. It was also stated in the minutes that the bishops intended to reach out to the Ministry of Ecclesiastical Affairs to suggest a working group be established and that, in their opinion, the issue should be part of future meetings at all levels of the Church. This was, however, according to *them*. Because, as a former Minister of Ecclesiastical Affairs stated in response to a bishop's engagement with the matter on an earlier occasion: "What ten bishops say in chorus is equal to the chorus of ten greengrocers" (Rønn Hornbech 2019). Hornbech's statement captures a common understanding of the People's Church as not built around hierarchical authority but rather a "well-ordered anarchy." The Danish People's Church, I was told by my interlocutors in the Church, is not organized in such a way that allows any one person, office or organ to make decisions on behalf of the whole. Decision-making, as well as the distribution of resources, is achieved through a decentralized structure along several different institutional axes organized around the parishes, the deaneries and the dioceses (Andersen et al. 2012). Hornbech's statement also summarizes the gist of the critique that was mounted on the bishops from many quarters within the Church after their announcement: given the People's Church is not organized hierarchically, the bishops did not have the authority to speak on behalf of the entire Church. In fact, as I would learn as the controversy around the bishops evolved, *no one did*.

When I began my PhD research in 2019, I did not know very much about the Danish People's Church. I had set out to ethnographically study what happened as a demand and desire to respond to climate change "travelled" into the Church, an institution that in many ways is the guardian of tradition and continuity in Danish society. My initial idea was to contact some kind of head office or a person in a central position to inquire about where in the Church I could find engagements with the topic of climate change. But as I looked for a way to contact the Church, it started to dawn on me that the Church as an institution was quite different from what I had expected. The Danish People's Church, I learned, cannot be approached as an institution in which someone has a full overview of what is going on. There is no head office and no position from where to gain an overview. On the contrary, one of the main tenets of the Church seems to be that no one should be able to know (and thus control) what is going on in its 2,340 parishes. This peculiar form of Church organization is something that I heard described in various ways: a "well-ordered anarchy" (*et velordnet anarki*); "the shapeless Church" (*den konturløse kirke*), implying an open and unbounded entity; a Church that "has five million faces – at least" (referring to the approximate number of Danish citizens); "a byzantine labyrinth" (*en byzantisk labyrint*), referring to how difficult it is to figure out who holds authority and/or responsibility on specific issues; and, poetically, that "it is the parish that breathes life into the Church." These descriptions show, among other things, how the Danish People's Church has a strong tradition of being ruled locally and "from below;" that it is the Church members – today approximately 71.4% of the Danish population – and the parish councils, rather than the clergy, the State or any other single authority (Rasmussen 2021). They emphasize that decision-making is decentralized, hinting at an egalitarian or at least anti-hierarchical, ethos (as also argued by Iversen 2019). That is, bishops are no more important than greengrocers and, therefore, what they think of the national 70% reduction goal does not represent the views of other Church members.

There are theological reasons for this emphasis on the Church as a decentralized, non-hierarchical organization. The Danish Church is an Evangelical-Lutheran Church, and it builds on the Lutheran idea of "universal priesthood" (Iversen et al. 1996). Luther is often quoted by Church actors for stating that "anyone who comes crawling out of baptism has been consecrated as priest and bishop." That is, all (baptized) individuals are equally sinful, equally redeemed and equally submitted to God's rule. Within the Danish Church, the clergy therefore hold intellectual authority within certain domains, but they are not elevated over other individual believers, and they hold no formal power over them (ibid.). Priests are understood – emphatically – as *serving* the congregation, while bishops have an *administrative* and *overseeing* position. Neither are positions of rule nor positions from which coercive power can be enacted. Hence, when the ten bishops made a statement about how they thought the Church should support the reduction goal set by the national government and were met by the view that "a choir of 10 bishops equals that of 10 greengrocers," the critique was informed by this Lutheran take on egalitarianism as a guiding principle of Church organization. When it comes to what the Church ought to do, the opinion of bishops matters as much as any other member's.

It is, however, not only the theologically informed egalitarian principle of universal priesthood that members alluded to when arguing for the preservation of decentralized organization. The egalitarian imagination also has another, albeit related, source: much more often it is evoked with reference to the values of "spaciousness" (*rummelighed*) and "religious freedom" (*trosfrihed*).

In the following, I scrutinize the cultural understandings embedded in the values of spaciousness and religious freedom and how they influence the organizational ethos of the Church. This exposition serves the purpose of exploring what underpins the statement that "ten bishops are equal to ten greengrocers" and why it seemed so impossible for the Church to commit to something like the national reduction goals in 2020. What underpins it, I argue, is a certain valuing of egalitarianism and of difference, both captured by the term spaciousness.

Making and taking room

In Danish, the term *rummelig* is used to describe the physical dimensions of material spaces, but it can also be used figuratively to describe a desirable social attitude and sociality in which there is "made room for everyone" (Anderson 2003), as well as to describe the virtue of being able to "accommodate" (*rumme*) others. *Rummelighed*, which can be directly, albeit clumsily, translated into "spaciousness," here denotes a cultural value (Robbins 2018) pertaining to social behaviour as well as an attitude. More than a comment on simply being spacious, the Church is also described as *valuing* spaciousness (Dylander 2023). It is thus a normative ideal – a value to hold and strive for in both principle and practice. It is associated with terms such as "inclusivity" and "tolerance." In the Church's self-understanding, "spaciousness" is a key value that captures its desire to be a broadly inclusive Church (Rubow and Engdahl-Hansen 2015). This implies broadness both in terms of types and classes of People and in terms of political convictions and theological variations. It is tied to the constitutional – and Lutheran – principle of *trosfrihed*, which can be translated into "religious freedom," but which more accurately describes the "freedom to believe" (whatever one wants). The broad consensus is that these values ought to define the organizational foundation of a Church that is supposed to be the

Church for the entire Danish People. That the Danish Church is called the *Folkekirke*, which is a compound noun consisting of "people" and "church," is meant to highlight that the Danish Church is not just a Lutheran-Evangelical Church but that it is also the Church of the Danish *People* (Hall, Korsgaard, and Pedersen 2015) and, importantly, not of the Danish *State* (Dabelsteen 2012). The principles of spaciousness and religious freedom have underpinned the organization of the Church since it was established, and they are widely understood to be just as relevant – if not more so – today. For example, in the document that expressed the newly elected government's vision for Danish society in 2022, it was stated that:

> Denmark is a Christian country, and the Danish Evangelical Lutheran Church holds a special status as the People's Church. The Government will sustain this special status. We want a church for the people that is based on freedom [frihed], equality [lighed] and spaciousness [rummelighed]. At the same time, it is crucial for the government that there must be freedom to believe in what one wants, as long as it is done with full respect for the right of others to do the same.
> (Statsministeriet 2022) (author's translation)

While Church actors argued that this could be taken to mean many different things – and therefore potentially nothing (Dylander and Lützen Ank 2022) – there was (and is) agreement on how these values imply that no one should have to submit to a narrow theological doctrine or a political ideology within the Danish Church and that there should "be room for" (*være plads til*) each member or congregation to form their own opinions and beliefs – albeit within the bounds of the broad tradition of Lutheran Evangelicalism and the specific creeds and confessions that the Church builds on (Gade 2022). Hence, when controversies play out in the Church around matters such as same-sex marriage or the ordination of female pastors, the virtue of spaciousness is evoked as an organizational principle in favour of arguing that there ought to be rooms for pastors in the Church who are against both of those things. The Church can only be spacious if pastors and congregations are allowed to differ on such matters.

In this vision of the Church's organization, there is an emphasis on the combination of individual freedom and collective commitment, heterogeneity and unity and freedom within the bounds. This particular understanding of egalitarianism entails a resistance towards being (forced into being) too much alike. This differs from the otherwise dominant analytical framework for understanding Scandinavian egalitarianism, namely "equality as sameness," as it was famously coined by Norwegian anthropologist Marianne Gullestad in the early 1990s. The gist of Gullestad's argument was that in Scandinavian countries, equality is less thought of in terms of equal opportunities or equal freedoms, but more in terms of the quality of being "alike" (ibid.). The equality that Scandinavian cultures value and practise is based on a culture of avoiding differences and stressing similarities, Gullestad argued (ibid.). The case of egalitarianism in the Danish Church, bound as it is with the value of spaciousness, seems to be a bit different. Here, equality is about avoiding making things – people and beliefs – the same. Actually, equality and freedom seem to be confirmed when disagreement is allowed to arise, as this is a sign of religious freedom, which again is a sign of no one in the Church being able to dominate others (see Høgh 2023). Hence, equality is preserved as differences are evoked and accepted – even if gender discrimination is accepted.

As the logic of carbon footprinting travelled into the specific socio-cultural context of the Danish Church, the values of spaciousness, egalitarianism and religious freedom were implicitly and explicitly used in the service of rejecting the kind of engagement with climate change that the bishops suggested. When the Church was encouraged from several sides to support – as a unity – the national reduction goals, the rejection of doing so was posed as a critique of what Church actors understood as an attempt to streamline the Church and eradicate the freedom of its members. To be spacious and to protect the freedom of Church members, each congregation should be able to respond to the climate crisis in whatever way most suits them.

This solution, however, was not completely satisfying to the newly elected Danish government in 2020, which had been elected on – among other things – promises of an ambitious climate strategy. As I describe in the next section, political pressure was put on the Church to develop collective responses to the climate crisis if it wanted to avoid governmental interference.

The State and its powers

In January 2020 – a few days after the bishops' statement – the left-wing political party Alternativet proposed the passing of a new law that would oblige the Church to lease its land exclusively to organic farming or reforestation projects. Each parish owns significant tracts of land, which the parish councils often lease to local farmers to earn money for Church activities. The proposition made by Alternativet entailed that the Parliament, as the supreme legislative body of Danish society, would regulate this matter, allowing parish councils to make use of their land exclusively in the aforementioned ways. This proposition evoked as much an outcry as the suggestion made by the bishops, as it was thought to violate the principle of keeping an arm's length distance between the State and the Church in order to secure the freedom of the latter. As one bishop stated in an interview in the newspaper Christian Daily: "I think that, fundamentally, it is a good idea that the land is farmed organically, but it should be something that the local parishes decide themselves. It shouldn't be forced on them from the National Government" (Hein 2021). In other words, Church actors may agree with the goals of politicians, but they might not agree with the methods they propose. If the State began to legislate on the basis of political interests, this would compromise both the spaciousness and religious freedom of the Church's members, the argument went.

But while the idea of the State enforcing such a law on the Church caused controversy, it was not unconstitutional. The Parliament *can* pass laws that apply to the Church and has done so in the past (Christoffersen 2012). But how it may do so – and whether it should – is a source of ongoing debate, as it is actually rather unclear to most Church actors (and the Danish population at large) what the relationship is between the State and the Church. The Constitutional Act of 1849 states that "the constitution for the Danish People's Church should be decided by law." This is known as a promissory clause (*løfteparagraf*), which indicates that an agreement on the shape of such a constitution had not yet been reached at the time of crafting Denmark's Constitutional Act (Holm 2012). The paragraph from 1849 is still there, unchanged, which shows that – even if most Danish citizens are unaware of it – the relationship between the Church and the State remains principled (albeit not formally) unresolved (Christoffersen 2012). Who governs within the Church is not clear, and until it becomes settled, Parliament has the legislative power to regulate (certain) matters within the Church. But the unresolved nature of the Church's constitution creates much disagreement about the scope of the State's political and legislative command over Church matters and what an appropriate use of this might be (Christoffersen et al. 2012). On the question of how far into the Church the State ought to reach, a particular *tradition* of governance has been established, my interlocutors would tell me. Namely, that of non-governance, or of the State not asserting itself. In an official ministerial memorandum, the State's reticence to legislate on Church matters is defined as a "constitutional custom" (*forfatningssædvane*).

During the period of my fieldwork, the then Minister of Ecclesiastical Affairs explicitly referred to this tradition in connection with a controversy around whether the Church should allow congregations to discriminate against female pastors:

> Personally, I have no doubts about my opinion on that: one should not discriminate on the basis of gender. But to say that in interviews or other places where I am asked to respond to the matter as the Minister of Ecclesiastical Affairs… Normally, of course, a minister would try to establish a political mandate to legislate based on his or her political opinion about a matter. But not in regard to the Church. Here you ought to respect the tradition of making decisions based on dialogue and in collaboration with one another. In keeping with tradition, it is to a high degree the Church itself that must take the necessary steps forward.
> (Danish Broadcasting Corporation 2022) (author's translation)

The point is that the Minister of Ecclesiastical Affairs does not decide very much, even if she or he has the legislative power to do so. The Minister is expected to listen to the various Church actors and to enter into dialogue with them on issues that concern them – but she should only intervene on the initiative of the Church itself. While Parliament can, in principle, make far-reaching decisions on behalf of the Church, it does not do so due to a "tradition" of respecting religious freedom as well as its sovereignty – even though this sovereignty, as we learned from the controversy around the bishops, cannot be pinned to any particular position within the institution.

That the Church is in principle, but not in practice, governed by the State safeguards the values of decentralization and egalitarianism and ensures that no one can claim coercive power. In other words, it is *because* the Church is a (kind of) State-Church that it can remain egalitarian. In Denmark, the State is, one might say, the guardian angel of egalitarianism – not only between individuals, as God and Luther have it, but also institutionally, since no Christian clique or faction can overrule another. If the Church was "freed" from the State, it would have to establish its own governing organ – a synod, for example. In such a constellation – which has been realized in other countries, for example in Sweden – one runs the risk, my interlocutors would say, of losing that cherished spaciousness and religious freedom.

Animating the well-ordered anarchy from above

It seems that the "ten bishops equal ten greengrocers" argument could have been what put an end to both the controversy spurred by the bishops and that by Alternativet. The Church could have avoided involvement in the climate agenda and the nightmare of conceiving a process for how to act without forcing any single parish, member or pastor to submit to anyone else. But here the State suddenly raised its head. It would not leave the Church be. Although Alternativet's proposition was rejected by politicians and Church actors alike, it was interpreted by the people I met in the Church as a looming threat. Following the processing of the proposition in Parliament, the then Minister of Ecclesiastical Affairs put pressure on the Church to develop collective solutions if it wanted to avoid governmental interference. This pressure was not communicated to the public, but, according to my interlocutors in the Church, it was felt in various meetings between the Minister and the different Church organizations throughout 2020. As one representative from the National Association of Parish Councils told me when I interviewed him about the matter, the Minister – as well as other politicians – was "putting feelers out in the Church" to find out whether Church actors intended to do something about the matter themselves or if regulative interventions would become necessary. This interpretation of the situation was shared by most Church actors I spoke to: the political system would stay out of the matter on the condition that the Church figure out its own way of doing *something*. This was the State displaying its mandate to govern.

This threat evoked two interrelated nightmares for the Church – that of state intervention, of manifesting a hierarchy between the State and the Church, *and* that of establishing an internal hierarchy and authority to mobilize the Church as a unity. Hence, the threat of law evoked a kind of catch-22 situation for the Church: to keep its independence from the State *and* protect the ideal of spaciousness by keeping internal leaders at bay, its members had to act as a collective body, a unity. The problem the Church confronted, then, was how it could possibly respond to climate change as *the* Church, showing itself to be a responsible societal actor, if no one was authorized to articulate a position on its behalf. As we see in the next section, a response on behalf of the entire Church was formulated and institutionalized in late 2021 – even if the controversies in 2020 made it seem as though such a response was an institutional impossibility. This response hinged on the logic of carbon footprinting.

The green transition of the People's Church

The threat of state intervention stirred the oceans of the People's Church during my fieldwork, and in late 2021, the Church announced its official response to climate change. This took the form of a collaboration between the bishops, the National Association of Parish Councils and the Association of Deans, who together launched the green transition of the People's Church project (*Folkekirkens Grønne Omstilling* – from here on the FGO). This project – and the collaboration behind it – was based on a programme detailed in a 12-page document stating that the Church as a "unified organization" would pursue its commitment to contributing to "the green transition of society" (FGO 2021). The project was planned to run for four years, from 2021 to 2025, and its concrete goals were that by 2025, all parishes and deaneries would have formulated their own local "green" strategies; that tools, guides and inspirational materials would be developed for the parishes and deaneries to use in the realization of such strategies; that legal regulations hindering or delaying sustainable initiatives would be adjusted; and that a mapping of the Church's overall carbon footprint had been undertaken at the beginning to form a baseline so that, when it was remapped at the end of the project period, the results of the work could be measured.

It is the last goal that this chapter is particularly concerned with. Given that the FGO did not, in fact, have a mandate to make anyone within the Church do anything, it seems that the success of the project relied on these measurements and the reductions they would hopefully display. Although all relevant Church actors – lay and clergy – were carefully represented in the programme's steering board through their democratically elected representative bodies (to ensure that it was neither one nor the other who spoke on behalf of anyone), the project was not given a mandate to make any decisions on behalf of the Church or the individual parishes. Its mandate was simply to gather and disseminate knowledge, to inspire and to document the actions of those it sought to inspire. According to John, an MSc in Nature and Forest Management who was employed to manage the project in late 2021, the project was something akin to a "nudging campaign." "We can disseminate information, motivate and inspire," he said to me when I interviewed him in early 2022, "but we cannot force anyone to do anything." This was an important basis for the collaboration: initiative, action and funds ought to come from below.

The Church's official approach to green transitioning was thus not to make specific reduction goals and carve out a clear path for how to reach them. Rather, it would pursue a "nudging campaign" that consisted of disseminating information to the parishes and deaneries and then hope that they would take the initiative to adopt measures that contributed to the green transition of society. The strategy was thus in line with the Church principle of decentralized decision-making. This, at least, was the intended *internal* effect of the FGO's approach. From the outset, the project's communication strategy was very careful to communicate that it sought to support initiatives taken at the local level of the Church. But the project was supposed to signal something different to the outside: namely that the Church, as a "unified organisation" (FGO 2021), was taking responsibility for its green transition. This was indeed what the State had demanded of it. But how, one may justifiably ask, can dissemination of information about possible measures that are already available to the public qualify as taking responsibility?

This is where carbon footprinting enters the stage. Because this, it seemed, relied on the FGO's ability to document that things were in fact being done in the Church that could positively contribute to the national project of reducing Denmark's carbon emissions. In the next section, I examine how the prospect of presenting a carbon footprint came to organize the response of the Church to climate change and how the production of such a footprint may in turn come to influence the Church's internal organization.

A technology of distance

FGO's programme document states that the project was informed by an

> "acknowledgment of the fact that the legitimacy of the People's Church in the context of national debates about green transitioning depends on the Church's ability to document and make visible local progress and results by way of joint statistics, knowledge and communication."
>
> (FGO 2021)

One important goal for the project was therefore to make the Church's contribution to achieving the national reduction goals visible not only in qualitative but also in quantitative terms. As an energy consultant specializing in church buildings told me at the beginning of my fieldwork, local initiatives to reduce energy use had been undertaken for many years in parishes and deaneries, but documents of such efforts were not archived anywhere; hence, these were unknown to both the public and the Ministry of Ecclesiastical Affairs. Such initiatives could not represent the Church as a whole, and that was, it seems, what Parliament was asking it to do. To account for the Church's overall reductions in carbon emissions would, however, show the responsibility taken by the Church as a unified organization. A big chunk of the 18.7m DKK allocated for the project was therefore devoted to mapping the Church's carbon footprint and to developing methods to continue monitoring emissions. Such monitoring could show the improvements that local Church actors had taken on their own initiative. The project management could "nudge" them to take such initiatives, and by measuring and monitoring carbon emissions, they could confirm that their nudging campaign (or other factors) influenced the Church to take responsibility for contributing to the "green transitioning of society" (ibid.). And, as we have learned, much was at stake in being able to document such an act of taking responsibility.

Like in much corporate carbon accounting (Lippert 2011), the role of calculating and displaying a carbon footprint seems to be two-fold here: it enables the Church to trace and document reductions *and* to communicate responsibility on behalf of the Church. These two functions of employing carbon footprint calculations in the Church's response to climate change were laid out to me by Søren, an energetic and charismatic dean in Eastern Jutland – a pioneer in counting carbon in the Church. While many parishes and Churches have had their energy consumption calculated and assessed in terms of kWh and amounts of fuel for their heating systems, by 2020 carbon had not yet been employed as a measuring stick. But in the spring of 2020, Søren carried out a comprehensive mapping of an entire deanery's carbon emissions – the first of its kind in the Church context.

I interviewed Søren a few months after the project was completed and a report had been launched. Søren explained how he had initially formed the idea of accounting for the deanery's carbon emissions when he heard one of the bishops explaining what the bishops had meant when they stated that the Church ought to contribute to the government's reduction goals:

> So I am there, right, taking a look at those bishops when they have to answer questions: 'Well, then, how do you plan to do this?' This moves us immediately into the empirical. You are pushed into having to deal with some numbers. You must consider some practical, concrete categories. And that immediately challenges the bishop. He is standing there, in front of Roskilde Cathedral on TV and he says something like 'Hmm … Well… I guess we can't very well isolate that… There are some issues with the architecture. And with cultural heritage perspectives…' So he stands there and actually becomes a bit meek. How can we actually do this? And that is, then, what I bring down here to where I'm sitting. Because I'm positioned somewhere else. I am positioned where the Church operates. (author's translation)

To Søren, it seemed like the bishops were simply "voicing good, but abstract intentions" that had no bearing on any facts. Søren wanted to bring the issue of climate change mitigation from the sphere of opinions and intentions to that of actions and practices. He wanted to take up the challenge of figuring out the pragmatics of how a reduction of carbon emissions could be achieved. What would have to be done? What were the facts? The job of a dean, he said to me, is about managing the practicalities of Church life – of taking action, not simply airing opinions. He thought the bishops were far removed from reality, and he wanted to prove that the Church – on a local, operational level – was genuinely able to do what the bishops could only dream of.

Søren thought the first step in achieving this would be to account for the current level of emissions, to create a baseline and to create indicators of where it made the most sense to direct their attention and efforts. In other words, to account for the deanery's current carbon footprint. For the efforts in the Church to be efficient, numbers and data were needed – "we have to know what we are doing," Søren told me. But he also described at some length – and in the genre of a thrilling story – how he gathered a group of passionate people in his deanery to develop the project; how they invited Church employers and employees – parish council members, gardeners and church tenders – for a big dinner ("you must not underestimate the importance of a well set table," he said) and on a veritable "road trip" to various parishes in the deanery to discuss the issue and bring everyone on board.

And then, about two-thirds into our interview, a fundamental shift took place in our conversation. From narrating the project's timeline, ambitions and manoeuvres, Søren suddenly told me that the entire story he had been telling me was, in fact, the one he told all the people in his deanery to engage them in the green agenda. He said that he, in fact, understood his own role as dean primarily as that of a "storyteller" – someone who could interpret what was going on and make sense of what courses of action should be taken and why. To make use of numbers and facts, he said, establishes legitimacy around the stories one tells and hence contributes to engaging people in them. He had therefore carefully and continuously narrated the project based on a plot that cast the deanery in its entirety and the individuals partaking in the project as "frontrunners" in the Church's green transition. It was a motivational story, a rhetorical trick, a sleight of hand. And this was the case not only with the story but with the whole carbon footprinting project as well. The carbon emission numbers were produced to do what the bishops were incapable of: establish legitimacy around disparate and individual greening efforts made within the Church, to show that the Church was serious. The carbon footprinting project was a strategic device meant to engage and organize people towards a common goal. On the one hand, it derived its legitimacy by drawing on what Søren referred to as "science." On the other, it motivated people by way of its narrative of the visionary deanery.

In our interview, Søren seemed to provide me with two logical strands pertaining to carbon footprinting. First, he showed how the Church could operate and not just "express intentions," arguing that knowing the facts and basing strategies upon them is superior to "just talking." Carbon footprinting equals qualified action. Second, he turned this idea of fact-based "operationalizability" – the ability to know and act rather than simply think and dream – into a device that could engage people. Basing the project on facts helped to establish it as sound and serious. To frame the project as such mobilized people to act.

Suddenly the carbon footprint project seemed to be *the alibi* for engaging people passionately in the green transition. Hence, the ambitious project of accounting for the deanery's carbon emissions was not (only) the practical means of achieving carbon reductions but also a means of engaging people in the climate issue. Søren had engaged his colleagues and parishioners in a project he found important not only by telling them that it was important, but also by positioning him and them in contrast to the Church's top officials, who purportedly do the talking but not the walking. The deanery was configured as a frontrunner by way of presenting the parishioners' efforts as a case of working from the facts.

Søren's sudden turn in our interview exhibits an emic theory about how numbers in the form of carbon footprints can be used strategically in communication. What Søren let me in on was that numbers do not only signify what they supposedly represent – in this case, the amount of carbon emissions – but that they also signify responsibility and genuine action in and of themselves. Indeed, Søren contrasted the use of measurements and numbers to the mere expression of good intentions that he ascribed to the bishops. In the theory of numbers presented by Søren, carbon numbers are "performative" (Lippert 2016): they enact environmental engagement simply by being evoked.

The deanery's report – which Søren called "a story supported by numbers" – became a key reference point for people who were in one way or another engaged in climate change mitigation within the Church. The deanery's project is mentioned several times in the documents and correspondence that led to the establishment of the FGO.[2] John, the FGO project manager, had a similar take on the purpose of displaying carbon emissions in the form of numbers in the Church's mitigation efforts (although he did not exactly say that they served as a "rhetorical trick"). It was vital, John said, that the project was able to show results that could be measured quantitatively and not just qualitatively, and that the results could be, literally, "nailed to the church door" (a Danish saying that implies a statement is the truth and can be defended as such). Being the largest civil organization, what the Church did and what guided its mitigation efforts should be "transparent and sound," as John said to me.

This perception of the function of numbers can be described by what Theodore Porter has called "mechanical objectivity" (Porter 1995). Answering his own question of why numbers have become such an authoritative and convincing device in science as well as in neoliberal government, Porter argues that it is because they purport to produce objectivity (ibid.). To produce, collect and display numbers, he writes, is based on strict, mechanical rule-following and is thus supposedly free from subjective interpretation: "It implies personal restraint. It means following the rules. Rules are a check on subjectivity: they should make it impossible for personal biases or preferences to affect the outcome of an investigation" (ibid. 4). Porter argues that mechanical objectivity has a powerful appeal to the public as it establishes a sense of accountability. Numbers can be trusted as, in contrast to politicians – or even scientists – they are disinterested (ibid.).

When John says that the Church's reduction efforts need to be framed in a manner that shows their results to be indisputable and trustworthy, his perception is based on the idea that numbers convey facts that are in fact objective. It is by displaying indisputable results that the Church can show that it is "more than talk," as Søren said. The numbers, then, serve as more than simply displaying supposedly objective facts about the world; they also communicate to the State and to the public that the Church takes its responsibility seriously. Furthermore, this appeal to mechanical objectivity serves the purpose of satisfying the State so that it will keep a proper distance from the Church. Porter's suggestion to think of numbers as "a technology of distance" (1995) takes on a quite literal meaning here: the FGO can be said to use numbers – in the form of carbon footprints – as a way of keeping the State at bay. According to Porter, numbers are such powerful communicative devices because they can be used to communicate across distances and differences (ibid. ix). This is due to the "strict rules" for collecting them and the way they convey results in "a familiar, standardized form," which makes it possible for someone who was not present to understand how something worked (ibid.). As such, numbers can function as a shared language between distant, communicating parties. This implies, Porter argues, that reliance on numbers minimizes the need for personal knowledge and trust (ibid.). Numbers are therefore not only a technology of distance in the sense that they overcome distance; it is also because they work across distances that they can be trusted (i.e. the communicating parties do not have to become more intimate in order to understand and trust each other).

In recognizing numbers as simultaneously overcoming and retaining distance, one might say that calculating and displaying a carbon footprint for the Church simultaneously creates a desired distance between the State and the Church *and* the necessary proximity between them to make the former accountable in the eyes of the latter. The numbers bring the dispersed climate actions of the Church close enough for the State to "see" them and trust that the Church will take responsibility on its own, thus rendering state intervention unnecessary. Hence, in the context of the Danish Church, the idea of numbers as a "technology of distance" can be used to describe how the Church deploys numbers as a way of keeping the State at a distance. However, something more is implied by Porter's suggestion to think of numbers as a technology of distance, namely the way numbers create distance between what they enumerate. In the next section, I attend to how this aspect of numbers as a strategy of *external* communication was also important for the Church in its grapplings with the *internal* problem of acting like a unity in a context of decentralization.

Divided by faith, unified in numbers

What I provided in the previous section was something like an account of the analyses of the function of carbon footprinting that my interlocutors themselves express. I would like to add an additional analysis of the effect of introducing the logic of carbon footprinting in the Church. Because carbon footprinting also did something else, which was, one might say, a prerequisite for representing the Church as a responsible societal actor: namely, that accounting for the Church's carbon footprint cast the Church as *one*. The Church can be said to leave *one* footprint even if it does not have one voice. Hence, while carbon footprinting serves the purposes of creating legitimacy, motivation and of holding the State at bay, it does so by achieving what is otherwise considered impossible: by representing the Church as a unity.

In the above section, I described how John and Søren agreed that scientific measurements are important for the Church because numbers are perceived to be devoid of subjective interpretations and indicative of a sound and operationalizable approach to climate change mitigation. This communicated that the Church was a responsible actor in the green transition of society. In many ways, this seems like a typical case of accountability being used as a means of neoliberal governance to prevent state regulation, corporations (or, indeed, churches) presenting themselves as self-motivated, responsible and accountable actors in the green transition (Lippert 2016). But what is remarkable about the case of the Danish Church is that there is, in fact, *no one to really hold accountable*. Or at least, this is the idea that the Church has been eager to protect: that no one can make decisions on behalf of the entire Church, and no one can be responsible for everything that goes on within it. The presentation of a carbon footprint, however, does seem to render the Church an entity that can be held responsible. The footprint, in other words, was accepted as a statement on behalf of the entire Church – something that no person, office, or law was supposed to be able to accomplish. Hence, it is worth attending more carefully to what was taking place when the use of carbon numbers became a way for the Church to represent itself as a responsible societal actor. How come carbon numbers were internally accepted as a way of aggregating and representing the Church? As we have learned, other ways of attempting to enrol the Church as an actor in Denmark's green transition had been unsuccessful, spurring accusations of compromising the religious freedom and spaciousness of the Church. This again, I suggest, has to do with the way numbers gloss over particularities and purport to be free of interpretation and value (Porter 1995).

Before 2020, the most dominant actor working with green transitioning in Danish Church contexts was the organization Green Church (*Grøn Kirke*). In my interview with John, he referred to Green Church as the FGO's predecessor of sorts. He called them "rubber boot theologians," casting them as frontrunners, yet also, perhaps, of too radical a kind to ever become mainstream. "You know, the rubber boot biologists of the 1970s were also those types who wore certain sweaters and smoked 'funny tobacco'," as he put it. The approach of this organization is based on theology. The working group behind the Green Church frames their initiatives in a Christian vocabulary of "stewardship" and "care for Creation," rendering the climate crisis not only a technological and economic crisis but a moral and spiritual one. According to John and other people I spoke to – including Green Church members themselves – this approach was the reason that in 2021, the Green Church had never managed to attract more than 10% of the parishes making up the Danish Church since its establishment in 2009. The aforementioned values of spaciousness and religious freedom, combined with a widespread reluctance towards anything that can be interpreted as a belief in "justification by works" (*gerningsretfærdighed*), caused many parish councils to reject the approach of the Green Church, deeming it both politicizing and moralizing (Schyberg, 2024).

The success of the FGO, John asserted, therefore depended on the *avoidance* of theology. "It might seem paradoxical," he said, "given that we are talking about the Church." But once something becomes a theological matter within the Church, he went on, conflicting interpretations proliferate, things begin to go around in circles and nothing can ever be decided upon. Hence, theology is conspicuously absent from the FGO's programme (FGO 2021). Hence, the way the FGO is framed as a thoroughly pragmatic and scientifically based project that deals with things that one can "make univocal conclusions about," as John said, implies that Church actors supposedly can gather around to support it without all their internal differences getting in the way. The FGO can evoke a collective motivation *towards* something, John said, making it possible "to actually get things done."

Porter's argument about numbers as a technology of distance (1995) can also be used to shed light on the function carbon footprint numbers had *internally* in the Church. While numbers created distance between the State and the Church, Porter also writes about the distance between numbers and the reality they are meant to represent. The bridging of distance that numbers enable, Porter argues, hinges on the creation of another kind of distance – namely, *a distance between the numbers themselves and that which they enumerate*. Numbers bridge the distance between communicating parties by way of being one step removed from the concrete, particular reality that they are used to describe. Very few numbers and quantitative expressions pretend to provide complete and accurate descriptions of the world; instead, they rather conveniently summarize complex relations (ibid. ix). Hence, besides working across and thus retaining a distance between communicating parties, numbers can also be said to create distance in the sense that they gloss over the particularities of the things counted (see also Espeland 1997; Dalsgaard 2013; Lippert 2018). So, while the representation of the world through numbers may produce *proximity* in terms of enabling knowledge to travel across distance, it also creates distance in that "it erases the local, the personal, and the particular" (Espeland 1997, p. 1107). As Espeland states, referring to Porter, "standardizing calculations make the characteristics of those creating and manipulating the numbers less salient, inserting distance between the numbers and their users" (ibid. 1108).

If there are (theological, political and organizational) disagreements around whether, how and on what basis the Church should perform a green transition, these differences do not interfere with the production and presentation of numbers relating to carbon footprints. Carbon footprint numbers do not in themselves express the various intentions and opinions of the individual parishes. Numbers about carbon emissions are perceived to be sufficiently distant from the intentions and interpretations of people (i.e. carbon emission reduction numbers do not in themselves express the intentions and opinions of the individual parishes); therefore, they are not seen as a threat to the individual parishes' right to have their own opinion. It is not only the mechanical objectivity that numbers purport to embody that serves the Church's purposes, but that numbers can also conceal as much as they can reveal.

In a Danish Church context, the use of numbers perceived and presented in this way is clever, because any occasion for interpretation will make differences appear and conflicts arise, as John described. The Church is not supposed to agree with itself, and there should always be room for different perspectives and different interpretations. Numbers that display the carbon footprint of the Church circumvent this resistance towards unanimity, as they do not, in fact, reveal anything about the opinions or motivation of the people in the parish councils who are supposed to produce the numbers – those whose practices cause the emissions in the first place. This method renders the green transition a thoroughly organizational and technical problem, not a moral or theological one. Hence, each parish can calculate their carbon emissions for different reasons, but the communicative strategy of providing a carbon footprint for the entire Church is not compromised by such internal variation in intentions. Carbon numbers conceal a great deal about what goes on in the Church in terms of taking responsibility for the green transition. But they do signal that the Church actively attends to its emissions.

On the one hand, this can be said to be the purpose of displaying the carbon footprint of the Church: the Church ought to show itself as a unified actor, and the carbon numbers should, as a first step, establish it as a responsible one. In other words, carbon footprinting makes the Church *one* and *responsible* in the same move. But what has not yet been touched upon in the Church or by the FGO is what happens when the Church's emissions are put on display for everyone to see. I end this chapter with a small speculation on what might happen when the carbon footprint is measured and displayed to the public.

An accountable Church?

In the late summer of 2023, a map of Denmark was displayed on national television: the map showed how big a territory the Church covered by way of its right to reject the construction of windmills in the proximity of Churches. Dioceses have been granted a special right and a responsibility to reject windmills near church buildings because churches are considered cultural heritage. Since this special right came to the public's attention in 2022, there has been much controversy around it (Bahn 2022). Assisted by the map, critics argued that by possessing this right, the Church became a stumbling block to the green transition. Indeed, the map showed that if all churches protested against nearby windmills, it would be close to impossible to realize the transition to wind energy that much of the Danish climate mitigation strategy is based on. The map was displayed and debated on primetime national television, and the role of the Church in the green transition became the object of public dispute. The Church was scorned for working against the green transition – if anything, as a public institution, it should be expected to support the national reduction efforts, debaters argued.

The case displayed that the public is not indifferent to the position of the Church in the context of the green transition. It also shows that the public's interest in the position of the Church in regard to green transitioning grows exponentially when data about it becomes publicly available. To be accounted for, or indeed accounting for oneself publicly, invites the gazes of others who might conduct a critical examination. This fact also informed one dean's reflections on the method of carbon footprinting when I interviewed him about the newly established FGO: what would come after it, he asked? The carbon footprinting might have been enough to "calm the waters" and avoid state intervention, but knowledge also produces new commitments and new responsibilities, he said. "It is all very well with simply stating abstract goals of reductions, and displaying these by way of carbon footprints," he added, "but we all know that, if the hoped-for reductions do not show themselves, we enter a new stage: finding someone to hold accountable," he said with a grave look upon his face.

As Judith Butler points out, "accountability" has a double meaning: it refers to both what it is possible for a subject to be "held responsible for" and to how a subject can make itself "intelligible" (2005). According to Butler, accountable subjects come into being through a relation to an Other who demands of them that they make themselves intelligible in moral terms. The Other's question about who you are, Butler writes, requires that you provide an account of yourself that is comprehensible to the Other. The account must therefore take a form that is not the subject's alone but that is recognizable to the Other. Hence, the way in which a subject can make itself known to an Other – and thus emerge as a morally accountable subject – is grounded in a "shared horizon of intelligibility" (ibid.) that determines which moral questions can be formulated and which moral judgments can be made. As a subject must provide an account of itself, it is the "shared horizon of intelligibility" that determines the moral codes that the subject can be judged in relation to – what it can be successfully held accountable for.[3]

Carbon emission numbers are the language that the National Danish Government employs as a means to be held accountable for the changes – the reductions – it has set out to achieve. Even if the Church does not commit to any specific reduction goals, when it provides an account of itself by way of a carbon footprint, such goals might nonetheless be implicated in that very method itself. By making itself intelligible to the State (and to the public) as a unified, responsible actor in the very same language of the State, it engages in the kind of moral subjectivation that Butler outlines: it provides an account of itself within a shared horizon of intelligibility, constituted by certain moral codes – in this case, pertaining to carbon emission reductions. That is, by accounting for itself in a language that the State recognizes, namely that of carbon footprint calculations, the Church opens itself up to the moral evaluation inherent in the method itself. The carbon footprint map may turn out not only to represent the Church's emissions but also incite demands on how to deal with those emissions in the future.

Writing about financial reporting and auditing, Keith Hoskin (1996) defines the practice of accountability as something that "engages the self insistently" (ibid. 265), because it demands that one not only describes past performances and present circumstances but also bases one's choices on future potentials (ibid.). Hoskins argues that this type of accountability practice tends to have deeper effects in the long run. Accountability practices tend to both display responsibility for actions taken and shape what actions can and will be taken at all (see Strathern 2000 for a similar point about auditing). It conflates what *is* with what *ought* to be, as Hoskins writes (Hoskins 1996, 270). Hence, while this remains speculative, the calculation of the Church's and the parishes' carbon footprints may very well influence the organization of the Church more than originally intended: the (carbon) account it provides of itself may in fact make demands on who that self becomes.

Conclusion

This chapter has shown how the prospect of carbon footprinting was used in the Danish Church's efforts to both communicate environmental responsibility and to protect "the well-ordered anarchy" of the Church, that is, its decentralized organizational structure, its heterogeneity in terms of (Christian) beliefs and political convictions and its independence from the State. As a way of accounting for itself, carbon footprinting renders the Church as a unified actor that takes responsibility for the green transition but without claiming anything about who or what the Church is, what it believes and/or what it does to mitigate climate change. While it is perhaps unsurprising that numbers serve as "a technology of distance" in terms of keeping the State (and the public) at bay – this is, one might say, ESG reporting in a nutshell. A more surprising function is that the numbers also served as a technology of distance in terms of keeping theology and the Church organization apart. As I have shown in this chapter, carbon footprinting was chosen as a preferred method to prevent theological discussions and conflicts around the issue of climate change. By way of numbers, the Church could present itself as a unified actor, despite the internal rifts and disagreements over whether the issue of climate change should be a matter of concern at all.

An additional and much less intended effect of the carbon footprint is that when it presents the Church as a unity (in order to keep the State at bay), such a footprint might also in fact enact such a unity and render it accountable and, indeed, governable. As a first step, the carbon footprint is meant to be a method of representation, but it may also produce what it represents in a certain way: a governable Church entity. Indeed, when the FGO finally launched their carbon footprint map in early 2024 (after my fieldwork had come to an end), the editor of the media site Kirke.dk wrote that the map had prospects that might make "more anarchistically inclined parish councils sit uncomfortably in their chairs" (Gade 2024). Because the carbon footprint is not only a novel way of representing the Church, it is also novel in the way it provides an overview of the Church and offers (at least potentially) detailed information about each of the parishes. Such a detailed – and even accessible – map of the Church has never existed before. To map the parishes' consumption "down to the smallest microscopic detail," as the editor wrote, "opens up entirely new possibilities for many other economic management and saving measures than those purely climate related." The anarchistically inclined parish councils will perhaps ask themselves: what comes next?

Acknowledgements

I am grateful to everyone in the Danish People's Church who took the time to talk to me and to teach me about the intricacies of the Church institution and about Church life in general. My doctoral research, which this chapter is based on, was made possible by the collaborative research project Socio-Cultural Carbon, housed at the IT University of Copenhagen, funded by the Danish Research Council and undertaken by Steffen Dalsgaard, Andy Lautrup, Ingmar Lippert and myself in 2019–2024. I wish to thank my fellow project participants as well as all participants in the Carbon-symposium in Copenhagen in the spring of 2022 for stimulating discussions around the cultural complexity of carbon and of climate change and for their helpful comments and input for this chapter.

Notes

1 In 2012, the Church of Norway committed to take measures to reduce their carbon footprint (Den Norske Kirke 2015), and in 2021, the Church of Norway Synod furthermore committed to becoming carbon neutral by 2030. In 2022, the General Synod of the Church of England approved a "Routemap" for how the Church of England can become carbon neutral by 2030 (Church of England 2022), and in 2019, the Swedish Church officially stated that it was committed to becoming carbon neutral by 2030 (Svenske kyrkan 2019).
2 I obtained all the documents and e-mail correspondence that involved the Ministry of Ecclesiastical Affairs.
3 I am aware that Butler is speaking about human subjects and not organizations. However, I find that Butler's image of how a subject that could potentially become many different versions of itself is called forth as a singular subject as it is asked to give an account of itself. In the same way, I suggest that the Church, which is in fact multiple in the sense of being made up of many different parishes and positions, is called forth as a singular entity as it suggests to account for itself by a single carbon footprint.

References

Andersen, Svend, Peter Christensen, Peter Garde, Peter Lodberg, and Anders Jørgensen, eds. 2012. *Kirkeforfatning. Kirkeretsantologi 2012*. Copenhagen: Anis: For Selskab for Kirkeret.

Anderson, Sally. 2003. "Bodying Forth a Room for Everybody: Inclusive Recreational Badminton in Copenhagen." In *Sport, Dance and Embodied Identities*, edited by Noel Dyck and Eduardo E. Archetti. Oxford: Berg.

Bahn, Martin. 2022. "Kirker bruger vetoret til at bekæmpe vindmøller." *Dagbladet Information*, February 1. https://www.information.dk/indland/2022/01/kirker-bruger-vetoret-bekaempe-vindmoeller

Blok, Anders. 2011. "Clash of the Eco-Sciences: Carbon Marketization, Environmental NGOs and Performativity as Politics." *Economy and Society* 40 (3): 451–76. https://doi.org/10.1080/03085147.2011.574422

Bramming, Torben. 2020. "Sognepræst: Kast ikke folkekirken på klimapolitikkens bål." *Kristeligt Dagblad*, January 14. https://www.kristeligt-dagblad.dk/debatindlaeg/sognepraest-kast-ikke-folkekirken-paa-klimapolitikkens-baal

Butler, Judith. 2005. *Giving an Account of Oneself*. New York: Fordham University Press.

Christoffersen, Lisbet. 2012. "Den Aktuelle Danske Religionsretlige Model." In *Fremtidens Danske Religionsmodel*, edited by Lisbet Christoffersen, Hans Raun Iversen, Niels Kærgård, and Margit Warburg, 239–258. Forlaget Anis.

Christoffersen, Lisbet, Hans Raun Iversen, Niels Kærgård, and Margit Warburg, eds. 2012. *Fremtidens Danske Religionsmodel*. Forlaget Anis.

Church of England. 2022. "Net Zero Carbon Routemap." *The Church of England*. https://www.churchofengland.org/about/environment-and-climate-change/net-zero-carbon-routemap

Dabelsteen, Hans Boas. 2012. "Staten, Folkekirken Og Dansk Sekularisme." In *Fremtidens Danske Religionsmodel*, edited by Lisbet Christoffersen, Hans Raun Iversen, Niels Kærgård, and Margit Warburg, 339–358. Forlaget Anis.

Dalsgaard, Steffen. 2013. "The Commensurability of Carbon: Making Value and Money of Climate Change." *HAU: Journal of Ethnographic Theory* 3 (1): 80–98. https://doi.org/10.14318/hau3.1.006

Danish Broadcasting Corporation, dir. 2022. "Ægteskabet mellem kirke og stat." *Tidsånd*. Danish Broadcasting Corporation. https://www.dr.dk/lyd/p1/tidsaand/tidsaand-2022/tidsaand-aegteskabet-mellem-kirke-og-stat-11042202357

Den Norske Kirke. 2015. "Klimamelding for Den Norske Kirke: Klimafotavtrykksanalyse 2012 Og Samhandlingsplan for Klima, Miljø Og Bærekraft 2015–2030." https://www.kirken.no/nn-NO/om-kirken/aktuelt/kirken-med-egne-klimamal/

Dylander, Viktor. 2023. "Den rummelige folkekirke er trængt af fejlagtig tro på frihed." *Kirke.dk*, October 27. https://www.kirke.dk/interview/den-rummelige-folkekirke-er-traengt-af-fejlagtig-tro-paa-frihed

Dylander, Viktor, and Heiner Lützen Ank. 2022. "Kritik: Regeringspapir er 'en gratis omgang' på kirkeområdet." *Kirke.dk*, December 14. https://www.kirke.dk/politik/kritik-regeringspapir-er-en-gratis-omgang-paa-kirkeomraadet

Espeland, Wendy Nelson. 1997. "Authority by the Numbers: Porter on Quantification, Discretion, and the Legitimation of Expertise." *Law & Social Inquiry* 22 (4): 1107–33. https://doi.org/10.1111/j.1747-4469.1997.tb01100.x

FGO. 2021. "Folkekirkens Grønne Omstilling – Programbeskrivelse." Obtained from the Ministry of Ecclesiastical Affairs at https://www.km.dk/fileadmin/share/Faellesfonden/Folkekirkens_groenne_omstilling_-_Programbeskrivelse_-_Bilag_til_OMP-ansoegning.pdf, on September 25, 2024.

Gade, Kåre. 2022. "Frihed, lighed og rummelighed." *Kirke.dk*, December 19. https://www.kirke.dk/analyse/frihed-lighed-og-rummelighed

Gade, Kåre. 2024. "Folkekirkens nye CO2-regnemaskine har potentiale til mere end blot klimatiltag." *Kirke.dk*, April 23. https://www.kirke.dk/kommentar/folkekirkens-nye-co2-regnemaskine-har-potentiale-til-mere-end-blot-klimatiltag

Hall, John A., Ove Korsgaard, and Ove K. Pedersen, eds. 2015. *Building the Nation, N.F.S. Grundtvig and Danish National Identity*. Montréal, Québec, Copenhagen: McGill-Queen's University Press & Djøf Publishing.

Hein, Niels. 2021. "Biskopper advarer regeringen: Gå ikke for vidt med grøn lov." *Kristeligt Dagblad*, January 16. https://www.kristeligt-dagblad.dk/kirke-tro/biskopper-advarer-regeringen-gaa-ikke-vidt-med-groen-lov

Høgh, Marie. 2023. "Åndsfriheden i folkekirken er under pres." *Grundtvigsk Forum* (blog). https://grundtvigskforum.dk/viden/åndsfriheden-i-folkekirken-er-under-pres

Holm, Anders. 2012. "Mellem Frihed og Orden". In *Fremtidens Danske Religionsmodel*, edited by Lisbet Christoffersen, Hans Raun Iversen, Niels Kærgård, and Margit Warburg, 171–182. Forlaget Anis.

Hoskin, Keith. 1996. "The 'Awful Idea of Accountability': Inscribing People into the Measurement of Objects." In *Accountability: Power, Ethos and Technologies of Managing*, edited by Rolland Munro and Jan Mouritsen. London: International Thomson Business Press.

Iversen, Hans Raun. 2019. "Secular Lutheranism as Common Institutional and Mental Background for the Danish Welfare State and the Danish Folk Church." In *Individualisation, Marketisation and Social Capital in a Cultural Institution: The Case of the Danish Folk Church*, edited by Lisbet Christoffersen, Niels Kærgård, Margit Warburg, and Hans Raun Iversen, 65–86. University of Southern Denmark Studies in History and Social Sciences. Odense: Syddansk Universitetsforlag.

Iversen, Hans Raun, Lisbet Christoffersen, Theodor Jørgensen, and Preben Espersen, eds. 1996. *Det Almindelige Præstedømme Og Det Folkekirkelige Demokrati*. 1sted. Forlaget Anis.

Lippert, Ingmar. 2011. "Extended Carbon Cognition as a Machine." *Computational Culture* 1 (November): 1–17.

Lippert, Ingmar. 2015. "Environment as Datascape: Enacting Emission Realities in Corporate Carbon Accounting." *Geoforum* 66 (November): 126–35. https://doi.org/10.1016/j.geoforum.2014.09.009.

Lippert, Ingmar. 2016. "Failing the Market, Failing Deliberative Democracy: How Scaling up Corporate Carbon Reporting Proliferates Information Asymmetries." *Big Data & Society* 3 (2): 1–13. https://doi.org/10.1177/2053951716673390

Lippert, Ingmar. 2018. "On Not Muddling Lunches and Flights". *Science & Technology Studies* (November): 52–74. https://doi.org/10.23987/sts.66209

Lovell, Heather, and Donald MacKenzie. 2011. "Accounting for Carbon: The Role of Accounting Professional Organisations in Governing Climate Change." *Antipode* 43 (3): 704–30. https://doi.org/10.1111/j.1467-8330.2011.00883.x

MacKenzie, Donald. 2009. "Making Things the Same: Gases, Emission Rights and the Politics of Carbon Markets." *Accounting, Organizations and Society* 34 (3–4): 440–55. https://doi.org/10.1016/j.aos.2008.02.004

Porter, Theodore M. 1995. *Trust in Numbers: The Pursuit of Objectivity in Science and Public Life*. Princeton University Press.

Rasmussen, Jens. 2021. "Strukturdebat: Forholdet Mellem Stat Og Kirke." *Tidsskriftet Fønix* 2021 (3): 61–95.

Robbins, Joel. 2018. "Where in the World Are Values?: Exemplarity, Morality, and Social Process." In *Recovering the Human Subject*, edited by James Laidlaw, Barbara Bodenhorn, and Martin Holbraad, 1st ed., 174–92. Cambridge University Press. https://doi.org/10.1017/9781108605007.009

Hornbech, Birte Rønn. 2019. "Folkekirken er ikke et politisk parti." *Jyllandsposten*, September 27. https://jyllands-posten.dk/debat/blogs/birtheroennhornbech/ECE11641689/folkekirken-er-ikke-et-politisk-parti/

Rubow, Cecilie, and Anita Engdahl-Hansen. 2015. "The Wedding Horseshoe Enacted as the Devil, as the Trivial and as Blessing Machine." In *Between Magic and Rationality: On the Limits of Reason in the Modern World*, edited by Vibeke Steffen, Steffen Jöhncke, and Kirsten Marie Raahauge. Copenhagen: Museum Tusculanum Press.

Schyberg, Katinka Amalie. 2024. *Church and Climate Change in Counterpoint: An Ethnography of Environmental Engagements within the Danish People's Church*. Ph.D. thesis, Copenhagen, Denmark: IT University of Copenhagen.

Hansen, Mette Skov. 2020. "Biskop føler sig misforstået: Vi har ikke et klimamål for folkekirken." *Kristeligt Dagblad*, January 18. https://www.kristeligt-dagblad.dk/kirke-tro/biskop-beklager-misforstaaelse-vi-har-ikke-et-klimamaal-folkekirken

Statsministeriet. 2022. "Regeringsgrundlag 2022." https://www.stm.dk/statsministeriet/publikationer/regeringsgrundlag-2022/

Strathern, Marilyn. 2000. *Audit Cultures: Anthropological Studies in Accountability, Ethics and the Academy*. 1st ed. London: Routledge.

Svenske kyrkan. 2019. "Svenska kyrkans färdplan för klimatet." Last accessed August 21, 2024. https://www.svenskakyrkan.se/klimat

5 Carbon in Chinese notions of ecological civilization
Policy of quantification or philosophy of promise?

Charlotte Bruckermann

Introduction

The poster (figure 5.1) adorned a wall running along the Min riverside in Fuzhou, the capital of Fujian, the province where current President Xi Jinping first pursued his goal of creating an "ecological civilization" (*shengtai wenming*) as provincial governor (1999–2002). The poster encapsulates the vision the state promotes in propaganda campaigns surrounding the concept. The previous evolutionary stage of "industrial civilization" (*gongye wenming*), characterized by its polluting smokestacks, fossil-fuelled transport systems and heavy industries breathing fire into grey skies, should be surpassed by the ordered rural-urban landscape of "ecological civilization". In the future low-carbon city, the majority of the human population will live in glossy high-rises, nestled within a vast landscape of sun-soaked mountains, forests and pastures providing green energy, water and even air, as butterflies soar up into the blue sky.

As the intersecting crises of air pollution and climate change gain pace, environmental governance increasingly relies on the metric of "carbon" (*tan*) to address the ensuing challenges. The Chinese character 碳 *tan* refers to the chemical element of carbon, including the carbon dioxide molecule CO_2 (*er yang hua tan*, literally two-oxygen-combined-carbon). It also functions as shorthand for greenhouse gas emissions, for instance in state policies to "reduce carbon emissions" (*tan jian pai*) and official propaganda calling on citizens to lead "low carbon lives" (*di tan shenghuo*). The carbon metric most frequently emerges in connection with markets and finance, for example, in "carbon emissions markets" (*tan paifang jiaoyi*) and "carbon sinks" (*tan hui*), in its commodified and objectified form.

More concretely, as severe smog episodes and their associated "costs" rise in China, so do citizens demand for cleaner air, potentially challenging the status quo. The Communist Party under Xi Jinping has responded to calls for environmental redress by setting ambitious carbon targets, thereby moving beyond the post-Maoist paradigm of economic growth. These new political commitments introduced ecological redemption into the state agenda through promises of a coming ecological civilization and simultaneously attempted to foster "ecological consciousness" (*shengtai yishi*) among citizens. Therefore, state

DOI: 10.4324/9781003478669-6

Figure 5.1 Poster for Fuzhou Eco-civilization campaign.
Poster caption: 绿色城市和谐家园: 共创生态手牵手，生态文明心连心
"Green City, Harmonious Homeland: Hand in hand to create our common ecology, heart to heart in building ecological civilization." (author's translation)
Source: Photograph by author.

discourse on ecological civilization does not only foresee changes to the environment but also that the population incorporates a new form of subjectivity that is adapted to the coming era of environmental redress, or, in government parlance, a state of harmony between "humans and nature" (*renyuziran*).

In this chapter, I explore the work and practices of Chinese workers who are supposedly at the forefront of implementing carbon policies and philosophies. I situate their efforts within the socio-technical imaginaries and material-ecological environments they engage with to make their actions legible and meaningful. I follow investigations into how economists, politicians and data managers (Callon 2009, Dalsgaard 2013, Lippert 2018), as well as environmental managers, activists and green workers (Blok 2011, Lippert 2015, Whitington 2020), engage in the creation of "homo carbonomicus" – the carbon-calculating subject (see Blok 2011) – or, in this case, question its calculative premise in favour of alternate environmental imaginaries. Beyond corporate boardrooms and accounting spreadsheets, environmental workers face contested ethics and competing values on the ground (Lippert, Krause, and Hartmann 2015). Disassembling the calculative abstractions of "carbon as datascape" (Lippert 2015) forces the *political* performativity of carbon – for instance, amid civil society groups – into the foreground (Blok 2011). Here, I show how workers in afforestation projects for carbon offsets in Northern China destabilize carbon's reductive logics by attaching non-quantifiable values to their environmental work, especially in the form of kinship ties and their attitudes towards state projects, as they strive to contribute to a greener future.

The research for this chapter is based on fieldwork I conducted in the People's Republic between 2016 and 2019. Methodologically, this research involved "following" or "tracing" the carbon as a relationally constituted and contextually situated object. In this chapter, I focus on the deserts of northern China where afforestation projects aim to combat desertification and climate change through carbon offsetting. The rural citizens planting the trees and shrubs in the dunes challenged the way their work was compensated and emphasized their livelihood struggles in the harsh environment they inhabit as key motivations guiding their fight against the encroaching sands.

I begin by outlining how the state discourse surrounding "ecological civilization" ties carbon logics to cultural values that become elevated by political promises to revive the ecological foundations of Chinese civilization. I then contextualize this policy turn within the Chinese Communist Party's attempts to fill the ideological vacuum left by Maoism, in particular by delivering well-being through economic growth. However, as the environmental costs of rapid development gain pace, the state increasingly champions ecological restitution as part of the coming good life. Through carbon limits, the state project enfolds citizens' contributions within a neo-socialist commitment to civilizational resilience.

Turning to concrete examples of the emerging carbon economy brings these issues to life in their cultural diversity. The discursive framing of an exemplary "model" for afforesting the desert in Wuwei Prefecture connects to ancient teachings and legends that are revalorized through a contemporary socialist rereading. Nonetheless, big tech and green increasingly captured concrete projects and even personal trajectories through carbon offset targets.

I highlight the personal trajectory of Zhang Laomo,[1] who went by the name of "Model Worker Zhang" (*laodong mofan*), an official title bestowed upon him by former President Hu Jintao during a ceremony honouring him and his fellow "servants of the people" in the Great Hall of the People on Tiananmen Square in Beijing in 2005. Zhang Laomo received this honour for leading afforestation efforts as part of his family and community's "Green Desert" project in the Wuwei District of Liangzhou in Northern China. I met Zhang during fieldwork in 2019. Aged 51 at the time, Zhang was a sturdy man with a weathered face and calloused hands resulting from his life-long commitment to planting trees in the desert. His dedication to combating desertification was evident as he organized grassroots afforestation drives. Zhang led his village work team, as well as visitors from nearby towns and cities, in efforts to make the desert green. From a young age, Zhang felt a calling to fight desertification, choosing manual labour over formal education. Despite considering himself "uncultured" (*meiyou wenhua*) due to limited schooling, Zhang drew inspiration from Chairman Mao's thinking. He displayed a Mao poster above his desk alongside the God of Wealth on his family altar, thereby combining his neo-socialist fervour and commitment to civilizational continuity. This perspective characterized Zhang's evaluation of his life working in the golden dunes.

Fieldwork with Model Worker Zhang's family and village revealed that the actual work of planting relied on forms of hardship and resilience that existed in tension with the abstract carbon calculus. Kinship and village sociality framed local environmental work in the following ways: first, through personal loss and ecological perseverance; second, through the importance of (neo) socialist production teams as a community of workers; and third, by challenging the lack of transparency in carbon funding in their own project and in state forestry farms, where suspicions over corruption arose.

Theoretically, carbon credits involve quantifying, monetizing and exchanging tokens related to various "carbon"-related activities, particularly the emission, reduction, avoidance or absorption of greenhouse gases in the atmosphere. However, the process of generating carbon credits as tokens of economic value simultaneously absorbs political, social and environmental values into its logic, often creating conflicts. As quantitative and qualitative values collide, intersect and connect, they give rise to "frontlines" (Kalb 2024) and "boundary struggles" (Fraser and Jaeggi 2018) where people position themselves within contested fields. More broadly, some values can encompass these contradictions, forming "ideal" or "civilizational" values that nonetheless reveal fractures and fissures (Kalb 2024). For instance, the politics of "ecological civilization" reveal political tensions surrounding the environmental impacts of capitalism, on the one hand, while capturing their values for development by the state, on the other.

Here, I will explore the intersection of ecological civilization and cultural notions embedded in public policy, specifically through the quantification of carbon as part of a socio-cultural imaginary for a greener future in China. Building on observations that carbon calculations cannot fully accommodate (unforeseen) externalities (e.g. environmental destruction or social costs), I argue that carbon nonetheless has the capacity to appropriate cultural values that provide its technological and bureaucratic coding with a civilizing force. In the following, I show how green workers skilfully draw on contradictions and fragmentations in the carbon equation to challenge its totalizing logic.

Carbon in the socio-technical imaginary of ecological civilization

As ambitions to reduce carbon emissions become embedded in particular socio-cultural environments, the notion of "carbon" surpasses its quantitative, technical and economic form as a metric and gets entangled with cultural, ethical and even religious values (see Schyberg, this volume). Even the "economic" act of monetizing carbon reductions bears various technical and financial promises that surpass its quantifiable form, especially in relation to its potential for delivering solutions to pollution and climate change.

The aesthetics of decarbonization encapsulates the persuasiveness of abstract representations that chart carbon calculations, particularly in definitive forms, graphs and numbers (see Strathern 2000). Yet, these renderings simultaneously harbour the infinite relationalities of carbon that can manifest in a singular location, or even within a particular person (Knox 2020). Politically, carbon governance reshapes state-citizen relations, which extend beyond the spheres of industry and finance to influence socio-cultural relations on the ground (Blok 2011). In this chapter, I analyse the labour of workers in carbon afforestation projects and demonstrate how a green workforce emerges that reconfigures the value of their green efforts in concert with, and sometimes in defiance of, carbon logics.

Economic framings of carbon involve measuring, monetizing and exchanging diverse processes and activities. Carbon thereby fuses economic calculations with ethical, social and environmental values, often in contradictory ways (see Dalsgaard in this volume). This juxtaposes economic value with a diversity of situated cultural, ethical and affective values (Graeber 2013). The fusing of abstract calculation and ethical deliberation involved in representing carbon as part of a datascape not only highlights carbon's contradictions but also its possibilities in overcoming mathematical inconsistencies in favour of green promises for a brighter future (Lippert 2018). Rather than viewing these two realms of calculative and ethical carbon as incommensurable, I have elsewhere argued that they generate a relational frontline (Kalb, 2024, Bruckermann, 2024a). Moreover, in this ethnography the tension between quantitative and qualitative values of carbon becomes sutured by encompassing values that serve as ideal templates for utopian redemption and civilizational futurity.

In this chapter, this "frontline" emerges in terms of political legitimacy for the Communist Party as the environmental effects of capitalist development take their toll. While the politics of a carbon-calculating subject have gained influence in areas where atmospheric pollution and climate change are prevalent among people's environmental concerns, such as urban and post-industrial settings, in the deserts of Northern China, afforestation workers dealt acutely and immediately with dwindling water tables and desertification (see Bruckermann 2024b). While afforestation workers linked these concerns in more abstract ways to carbon, they more readily adopted the philosophical and social dimensions of ecological civilization to validate their environmental work and ecological contributions.

Mapping ecological civilization

The Chinese notion of ecological civilization includes many dimensions, from low-carbon transport campaigns and real estate construction advertising to dry policy documentation and fiery political speeches. All of these aspects aim to make ecological civilization discourse omnipresent in civic life and thereby shape moral communities through institutionalizing its political power and ethical force. In the poster above, for instance, ecological civilization is linked to the notion of the homeland, a word with high affective resonances that connects the intimate notion of the family with the entire nation-state (see Bruckermann 2019).

The Chinese government's promise to bring about an "ecological civilization" seeks to bring the "natural environment" in line with the historical and epochal human narrative of "civilization." The concept of ecological civilization is key to understanding the Chinese Communist Party's outlook on environmental governance beyond its carbon pledges, both domestically and abroad. For instance, in the context of the COVID-delayed COP26 in 2021, President Xi made a surprising and promising announcement via video link to the UN General Assembly in New York. Speaking on behalf of the People's Republic, Xi outlined an ambitious goal for decarbonization: "We aim to have CO2 emissions peak before 2030 and achieve carbon neutrality before 2060."

While the announcement came as a surprise to the international community, the targets broadly aligned with the pace of transition to a low-carbon economy outlined in previous Five-Year-Plans of the centralized National Development and Reform Commission (see FYP 2016–2021) and even tightened the target timeframe for peak emissions in subsequent years (FYP 2021–2026). The Fourteenth Five-Year-Plan, for instance, set carbon intensity targets of 18%, an improvement of days with good air quality in cities to a total of 86.5% and forest coverage rates of 24.1%. This comprehensive policy document takes into account decarbonizing energy production, improving energy distribution and resource utilization, greening various economic sectors, developing the circular economy, bolstering the country's environmental legislation and leading international cooperation on climate change (UNDP 2021).

While all this may sound like an extremely *technical* undertaking from the perspective of government policy directives, the *social* motivation for decarbonization takes the central stage in many other popular forms of outreach. Mette Hansen, Hongtao Li and Rune Svarverud's (2018) interdisciplinary research defines ecological civilization as a "socio-technical imaginary" that promotes "a vision of a society characterized by ecologically sustainable modes of resource extraction, production, and trade, inhabited by environmentally conscious and responsible citizens." In official English translations, Xi Jinping's notion of ecological civilization is sometimes translated as "ecological progress" (Xinhua 2021), presumably to emphasize the CCP's long-standing developmental approach.

In the Chinese context, Stephan Feuchtwang and Mike Rowlands (2019) trace the Chinese state's civilizing mission to early 20th century calls for Chinese "spiritual" civilization to stay true to its cultural foundations while absorbing knowledge and appropriating tools from others, including Western "material" civilization. For the early Reform Era of the 1980s and 1990s, Susanne Brandtstädter (2000) shows how the social institutionalization of state power unfolded alongside the stabilizing forces of restraint and stable or civilized behaviour. Moreover, Nicholas Dynon (2008) argues that contemporary civilizing campaigns are effectively morality campaigns that aim to replace Maoist class antagonisms as the motor of history with a harmonizing vision of material and spiritual progress through the Communist Party leadership.

Given the current challenges of the climate crisis, ecological civilization discourse envisions bringing humanity and nature into harmony. President Xi Jinping has made the fusion of ecology and civilization a goal throughout his political career. For instance, as Xi worked his way up the ranks of the Communist Party in the densely forested Fujian Province (1985–2002), he already proposed the construction of an "ecological province" as governor in 2001 (Liang et al. 2017). Championing the concept of an ecological civilization alongside an understanding of nature as an invaluable resource and motor for green growth, Xi took up the idea of harmony between humanity and nature set by his predecessors, including Mao Zedong's dialectic of humanity and nature (*renyuziran*) and Hu Jintao's synthesis of social harmony (*shehui hexie*).

Beyond techno-bureaucratic planning, state formulations explicitly substantiate ecological civilization with concepts derived from Chinese philosophical, political and environmental history. Most famously, Pan Yue has been credited with both creating and popularizing the concept as minister in China's Ministry of Environmental Protection, where he acted as the deputy and party secretary for over a decade in the 2000s. He became known as "hurricane Pan" for his turbulent crackdowns on pollution, his institution of environmental impact assessments, his pioneer Green GDP plan and his fervent warnings about China's and the global environmental crisis (Ma 2016). However, Pan Yue's most notable contribution, according to one of China's foremost environmental commentators, Ma Tianjie (2016), was articulating "a kind of environmentalism that is so organically Chinese that it has become subconsciously rooted in the national narrative."

As the deepening market reforms since the 1980s decentralized Maoist authority and shifted the focus of Communist Party narratives from class antagonism to social order, state morality campaigns increasingly integrated religious principles with ecological awareness, promoting environmental stewardship as a moral duty (Dynon 2008). Explicitly connecting ecological civilization to the country's three dominant religious doctrines – Buddhism, Confucianism and Daoism – the concept is embedded in a philosophical framework that emphasizes cosmic unity, the interdependence of all living entities and a reverence for nature (Hansen, Li, Svarverud 2018). Within the civilizational confrontation of spiritual and material priorities, ecological crises are cast as the consequences of prioritizing wealth over ethical responsibilities, and a return to traditional values is advocated as a remedy (Dynon 2008).

This evolution of civilizational discourse also incorporates elements from Marxist historical materialism and positions the pursuit of ecological civilization as the latest phase in human development (Dynon 2008, Hansen, Li, Svarverud 2018). Moreover, it serves the Communist Party as continuing an ancient civilizing mission while legitimizing its rule in the modern era, moving beyond mere economic growth to a broader conception of societal progress (Ding 2020). As environmental concerns abound, the state must deliver on metrics beyond economic growth, thereby demonstrating environmental concern for a happier, healthier and greener future (Ding 2020).

Carbon provides an opportune vehicle for meeting these political demands due to its ability to quantify a diversity of activities and processes while also making them comparable across different scales, from the individual as a citizen to the nation-state and even global spheres (Dalsgaard 2013, Whitington 2016).

Decarbonization experiments and examples

Model projects to combat carbon emissions are frequently glossed over as "experiments" in global environmental governance (Dalsgaard 2022, Callon 2009). Carbon markets that put a price on emitting pollution emerge as "fictive" in the Polanyian sense, that is, that carbon is not a commodity produced for the market, but like land, labour and money, carbon becomes commodified through its operations (Lohman 2010). Among the economists' toolkits, market mechanisms, including carbon markets, have limitations because they are not the most effective in safeguarding and accessing public goods and services, and they continuously create new externalities and attenuate inequalities (Callon 2009). Advocates for carbon markets nonetheless hold on to the possibility that optimal market design and real-life experiments have the potential to "civilize" these markets (Callon 2009).

In China, regional carbon markets and attempts to contain dust and pollution through atmospheric interventions are designated "scientific experimental zones" *shidian*, often translated as "pilot zones" in English, to intentionally keep the outcome open-ended (Bruckermann 2023, Zee 2021). Framing carbon governance as "experimental" mobilizes logic from the natural sciences whereby the accumulation of knowledge (rather than the reduction of carbon emissions) can make a policy a success, and failure can be deferred into an uncertain future (Bruckermann 2023).

In the Chinese context, the legacy and prominence of "policy experiments" can be traced back to the agile responsiveness of Maoist guerrilla tactics in hostile and unknown terrain (Heilmann and Perry 2011). Following this strategy, the Chinese government built on local experiments with regional pilot schemes that effectively tested variables (such as allocating or auctioning credits, rules on banking and borrowing, as well as the inclusion of offsets) within these emission exchanges. Subsequently, the national carbon market started operating in 2021, with parameters based on what they had learned from the preceding decade of experimentation (see Bruckermann 2023). Successful regional markets – for instance, those with high rates of industry compliance and large and frequent trading volumes – were championed as models to follow.

The use of models in Chinese governance has long relied on showcasing examples of positive behaviour for people to follow, especially in technocratic experiments that are sanctioned or fostered by the state, such as productive Maoist workers or diligent imperial bureaucrats (Bakken 2000). These exemplars embody moral commitments and viable models for the good life and encourage citizens to lead by example rather than strictly adhering to a rule-based ethical code, a phenomenon not unique to the late socialist context in China (see Humphrey 1997).

During my fieldwork, the notion of "ecological civilization" often elicited criticism, while "ecological consciousness" and "low carbon life" received much more positive responses as goals to strive towards. This revealed a widespread acceptance of orienting personal behaviour and subjective consciousness towards the shared project of environmental redemption while remaining sceptical of official proclamations regarding the coming era of ecological redress. However, as the state experimented with carbon values, it mobilized a synthesizing policy tool that brought together red socialist tactics and green market aspirations through the use of models, both as positive examples to follow and as cautionary prototypes to avoid.

Despite the unifying aims of ecological civilization discourse, significant differences exist in the varied positionality of citizens across China's rural-urban divide in relation to green restoration. As mentioned above, low-carbon lifestyles formed a central tenet of urban governance. Urbanites, influenced by mainstream media, focussed on lifestyle changes, linking affluence with green living. Initiatives like green-city campaigns and recycling drives were widespread (Hoffman 2011). Nonetheless, reliance on top-down state initiatives was prevalent, and citizens saw the state as accountable for environmental clean-up (Wang 2017). In contrast, rural residents often aspired to improve their own livelihoods while confronting environmental pollution and ecological degradation as an everyday, even mundane and unspoken concern (see Lora-Wainwright 2013, Bruckermann 2020). Rural residents sometimes framed their commitment to ecological restoration as crucial for survival, while state investments sought to link ecological improvement and employment (Bruckermann 2020). Systemic issues like lax law enforcement often remained unaddressed, as rural residents face the "double burden" of historically degraded landscapes and the task of future environmental restoration (Lord 2020). Despite green workers shouldering much of the tasks of ecological restitution as part of an emerging global "eco-precariat" (see Neimark, Mahanty, Dressler and Hicks 2020), rural Chinese workers often expected the state to spearhead and coordinate environmental efforts and were disappointed when this responsibility was not met.

While the government promoted the idea that being an environmentally enlightened person involved the individual pursuit of low-carbon practices, it simultaneously elevated model cities, villages and even individuals from both the city and the countryside as worthy examples for the public to emulate on a mass scale.

The Green Great Wall: from socialism to civilization

I now turn to carbon forest projects that are located in the Wuwei region of Northern China, which the government and official media have celebrated as models for ecological civilization. I begin by looking at how state discourse framed afforestation work through socialist and philosophical continuities, and recently recast local efforts through the lens of carbon finance. I then examine how workers on the ground enact their green work and engage with carbon logic.

Minqin County in Wuwei prefecture lies along the Hexi corridor, a relatively flat and arable plain that stretches across North China. Historically, this plane permitted passage along the Northern Silk Road, and it continues to link East and Central Asia as part of the contemporary One Belt One Road initiative. Not long ago, the region was maligned as a hotbed for devastating sandstorms that wreaked havoc in the region and blanketed the streets as far as Beijing in a fine coat of golden dust. Due to these meteorological phenomena, the region became part of a cautionary tale against ecological degradation. For instance, when Premier Wen Jiabao visited in 2007, he proclaimed that Minqin should not become a second Lop Nur – a huge lake in nearby Xinjiang Province that was swallowed up by the Gobi Desert in the 1970s. To avoid this fate, Minqin has become a site for experiments in fighting desertification that has received increasing popular and scientific attention during times of climate crisis (see Zee 2021).

Most notably, the region forms part of a flagship green engineering project of China, the Three North Shelter Forest Program (*San Beifang Hulin*), more colloquially known as China's Green Great Wall (*Weida de Lüqiang*). The TNSFP is a megaproject of late socialist engineering, a vast human intervention on the landscape through the planting of forest strips or shelterbelts that act as windbreakers to fight the expanding Gobi and to help fix topsoil to the ground to prevent dust storms (see Zee 2021). The project began in 1978 in parallel with Deng Xiaoping's market reforms and is due to be completed in 2050 when the ecological fortification is forecast to stretch over 4500 km. Despite problems with groundwater shortage, some sites have successfully turned landscapes degraded by industrial agriculture under Maoism into sites for eco-tourism, forestry, renewable energy and sustainable agriculture.

During fieldwork in the golden dunes of Wuwei, the fight against the desert took the mundane form of planting shrubs and trees that can fix the shifting sandbanks (see also Zee 2021). As part of a sequence of arduous tasks, workers laid out grids to make stabilizing fixtures across the sand dunes, using anything from nylon meshing and straw weaving to chemical fixing. They then dug holes for small seedlings of hardy desert bushes and trees before passing buckets of water along a human chain across the landscape to quench the initial thirst of the new transplants and to help them secure their hold in the granular surface. Despite these many precautions and support for the seedlings, survival rates were low and depended on factors beyond the planters' control, such as rainfall, wind and extreme temperatures.

In the afforestation sites overseen by the local Forestry Bureau, I mostly encountered manual workers who were either middle-aged women, whose children often lived in urban areas and whose husbands worked far away, or senior citizens who had retired from other income-generating activities. Local foremen supervised these workers toiling in the harsh desert conditions and drove accompanying vehicles with supplies and water. Rest periods often involved sharing food and drink and, on cooler days, even dancing. In addition to these formal labour arrangements, the Wuwei region hosted various volunteer and grassroots projects for anti-desert afforestation (Bruckermann 2024b). These initiatives typically relied upon visiting volunteers from nearby urban areas to plant seedlings, often motivated by professional commitments or family connections to the land.

A small handful of farmers have also taken anti-desertification measures into their own hands. This was the case for a number of farmers turned afforestation workers who contributed to the Green Great Wall in Babusha in Gulang County, adjacent to Minqin in Wuwei Prefecture. President Xi Jinping visited the project on 21 August 2019 and praised the "Yugong spirit" he found there. Yugong refers to an ancient Chinese folktale of "the foolish old man who moved the mountains" (*Yugong yishan*). Yugong lived near a pair of mountains that obstructed his way home. Therefore, he sought to move the mountains with hoes and baskets. When questioned about this impossible task, Yugong acknowledged that it was impossible to finish in his lifetime, but through the hard work of his children and the perseverance of many generations to come, the mountains would one day be moved. The gods were so impressed with his tireless commitment that they decided to help Yugong by separating the mountains.

Xi could not have evoked a more fitting myth to illustrate how the older generations of Babusha were fighting for ecological civilization. The legend of Yugong makes its first appearance in the Liezi, a Daoist text from the 4th century BC, and was subsequently republished by the Confucian scholar Liu Xiang in the 1st century BC in the Garden of Stories. Famously, Mao Zedong reinterpreted the tale to demonstrate "a utopian belief in the potentiality of the community or the collective, where perseverance and audacity are able to transform what seems impossible" (Moner 2021). Following in his footsteps, Xi Jinping praised the Babusha project as

> the 'six old men' continue to carry forward the contemporary Yugong spirit, facing their difficulties without bowing down, daring to turn the desert into an oasis of enterprising spirit, and enduring hard work, and their long-term achievements in making the Great Green Wall indestructible.
>
> (CPC 2021, author's translation)

These six old men of Babusha began their afforestation efforts in the 1980s and then committed the rest of their lives to fighting the desert after a catastrophic sandstorm struck Gulang County on 5 May 1993. In half an hour the sandstorm engulfed the entire area, devastating crops, destroying houses and killing residents, including 23 children (Wang and Li 2019). The forestry initiative received some government allowances for afforestation but also relied on forest-friendly cash crops to make ends meet.

In 2009, the project developers founded the Babusha Afforestation Company and bid for greening projects. Their most successful bid occurred in 2018 when their first graduate recruit, the technician Chen Shujun, landed 10 million yuan (1.4 million USD) of funding from Ant Forest, a green finance mini-app created by the third-party payment service Alipay. This app offers users not just cash and credit, but carbon accounts. Based on "green activities" tracked in the app, users receive "energy points" that they accumulate in a digital tree nursery. Once their pixelated sapling has been nourished with sufficient energy points, users can convert their virtual shrub into an actual tree planted by afforestation workers, like those of Babusha in the deserts of Northern China.

In Babusha, the work of afforesting the desert was transformed from a grand socialist project infused with civilizational depth to a carbon project fuelled by high tech and green finance. Despite the shifts in motivation for afforesting the desert, media accounts and participant interviews (e.g. Wang and Li 2019) emphasized the continuity of workers' values, including the emphasis on survival of the landscape, safeguarding future generations and celebrating mutual labour in the pursuit of environmental imperatives. Moreover, these elements were shared with the neighbouring county of Minqin, where I now turn to the fieldwork I conducted with the Zhang family, who led a grassroots project for anti-desertification afforestation – the "Greening the Desert" project – the environmental experiment that I briefly introduced in the opening section above.

The production team

In mid-May, a day before rainfall was predicted, Model Worker Zhang, his wife Jin and I clambered into their four-wheel drive and drove through their hamlet, honking and picking up neighbours willing to do a day's work on the anti-desert afforestation project. Villagers threw bags of steamed buns and bottles of water into the cabin before clambering onto the back of the pick-up, where they secured their hats and scarves for protection against the sun and wind. Before bumping along the roads to the desert area, we made a final stop at Zhang Senior's home to collect an assortment of sand shovels and dozens of plastic sacks filled with seedlings of a hardy desert tree (*huabang*). All thirteen female and five male workers who joined us in the desert were neighbours or relatives of the Zhang family, and most worked in the same subdivision, a single production team, in their commune during Maoist times. This meant that they had cooperated for many decades in fields and sand dunes, pooling labour and resources.

This long-standing collaboration shaped a well-rehearsed rhythm as the workers moved across the hills in invisible lines, dividing tasks in unspoken agreement and absorbing differences in pace without missing a beat. As older couples moved more slowly, they drilled fewer holes into the surface and therefore dropped seedlings less frequently than their more youthful counterparts, allowing the entire group to move across the dunes in shifting, wavering lines, at a single pace. Zhang drove his pickup loaded with water, seedlings and supplies across the precipices of the dunes, stepping out of the vehicle to survey the landscape, scout out appropriate afforestation sites and direct the group. Occasionally, he would also grab a shovel and participate in making holes or carry an extra heavy bag of seedlings back and forth from the hills to the planters.

As mentioned earlier, the Wuwei residents were not terribly concerned with pollution, carbon emissions or even climate change per se. They were more concerned about the immediacies of change brought about by desertification. Although the vast ochre landscape appeared like an uncharted sea of sand, seemingly only interrupted by a lone silver government observation tower gleaming in the distant sunlight, Zhang recognized and knew the names of various landscape features across the hills and valleys of sand. The actual names of these locations revealed how recently this area had become inundated by the fine grains of sand, bearing witness to past decades when the landscape teemed with vegetation and features like Sorghum Ravine (*Gaolianggou*) and Garden Nest (*Yuanwowo*). Despite the submersion of these spots under a blanket of sand, Zhang relied on their names to navigate the landscape, both in deciding where to go and in tracking where the production team had been. This became particularly important in the days and weeks following afforestation when many of the seedlings needed to be watered repeatedly to survive unless it rained.

Water was key to survival, as groundwater had slowly receded in the area from 40 metres to 100 metres below the surface since the 1970s. Some types of vegetation had to be replaced with new kinds, such as the characteristic *suosuo* trees that no longer survive in the area because their roots cannot reach far enough underground. Humans also faced the challenges of water scarcity. The Chinese military visited the village during my stay to deepen the well. They spent a full week of constant ramming into the earth with a mechanical vertical drill. The pounding drum that shook the entire village day and night beat to a monotonous rhythm: Humanity. Will. Have. Water. As a resource, water was so essential, and its exploitation had such a direct impact on others using the well that access to it was never decollectivized. While villagers farmed the individual plots of land that the village committee had allocated them, they pooled the use and cost of water, with a single meter measuring usage for the whole community.

In addition to working in agriculture, many villagers commuted or migrated for work. However, when discussing desert afforestation, many claimed to prefer this work, even if it paid on the lower end of the daily labour wage at 100 RMB for the day. If villagers sold similarly hard, manual labour in Wuwei, they might make between 100 and 200 RMB. However, they considered the afforestation work more convenient, less stressful and at times even fun. Zhang did not chastise or egg the workers on, and lunchtime breaks were times for relaxing and sharing food and even included the occasional desert dance session. Some of the villagers even came to the desert to avoid the tedium and isolation of domestic work, while others emphasized the need to earn money in parallel with family commitments. However, most put the sociality of the work at the forefront of their motivations, and they compared the afforestation work favourably with doing similar work for state-owned forestry farms.

Informed by socialist motivations and kinship obligations, this afforestation work relied on decades of cooperation, where workers were attuned to each other's rhythms and intimately familiar with the labour process at hand. The emphasis on survival in a harsh environment and how their afforestation work would ensure the wellbeing of future generations became even more apparent in the Zhang family's intimate familial motivations for engaging in green labour, which I evoke in the following section.

Safeguarding a home to return to

Zhang's whole family was no stranger to difficulty; the senior generation had vivid childhood memories of the famine years of 1958–1961, when they claimed to have eaten every last wild herb and even the bones of the animals to stay alive. In 1964, at the age of 20, Zhang Senior dug wells for the Maoist commune to provide water for agriculture. Despite the gruelling work of excavating ten wells, he fondly remembered this time and the comradeship in his brigade. In 1974, Zhang Senior was invited to join the Communist Party, and both his sons and two of his three daughters have since followed in his footsteps, much to his pride and joy. To this day, he liked to blast the red songs from that Mao era over a megaphone he installed in a watchtower across the dunes his family has turned green, as he reminisced about the achievements of his life, his family, his comrades and his country.

The village committee allocated these desert lands to Zhang Senior's family in the 1980s when the commune was disbanded; his family was granted 30 mu (or two hectares) of agricultural land and 20 mu (approximately 1.3 hectares) of desert. Initially, they carried water to the desert on camels from the village four times a day in plastic canisters. The camels were replaced by a diesel four-wheel-drive vehicle in 2006. In the 1990s, the family moved out into an old courtyard caved into the sandy hillside (a *diwopu*) in the middle of the reclaimed green dunes to be closer to their work. Zang Senior also experimented with different methods and developed his own "sand shovel" (*shamo qiao*) to be able to quickly plant seedlings deep into the dunes. Zhang Senior was motivated by survival, by an impulse to save their homeland from the sands, and to secure a future for coming generations.

Model Worker Zhang and Jin's youngest son – Zang Senior's grandson – Little Zhang, died of a brain tumour when he was only 14 years old. In the middle of planting season, the family put aside concerns about his worsening headaches and planned to postpone a visit to the urban hospital until all the seedlings were planted. Little Zhang himself was an ardent enthusiast of anti-desertification afforestation, and the family only realized the severity of his situation when a stroke left him partially paralyzed. Despite rushing him off to the hospital for emergency treatment, he died within months. The family recounted that before he passed, Little Zhang nonetheless encouraged his family to continue their important ecological work, and he now lies buried in a grove in the afforested dunes not far from his grandparents' home. The intergenerational promise of a greener future therefore motivated the Zhang family to plant trees and demonstrated an emotional drive and ethical impulse that eclipsed monetary, calculative, or climate-related concerns for absorbing carbon.

Since Little Zhang's death, his family has tried to put their health before the trees. Nonetheless, the loss of a son and grandson to the anti-desertification works was doubly painful for Model Worker Zhang, who explained in tears that his biggest motivation for struggling against the encroaching sands came after his sons were born, as he wanted to improve their surroundings and environment. In contrast to the emotional force driving the Zhang family's work to ensure the survival of their home, the surrounding landscape and each other, the calculative logic of carbon funding rang hollow. In conversation, various family members explicitly engaged with the carbon metric and challenged its reductionism in ways that revealed more holistic understandings of their afforestation activities.

Contesting carbon

Jin, Zhang's wife, had also dedicated her life to keeping the desert at bay. As one of nine children born in the same commune as Zhang, she grew up in a family that also afforested the desert, but in a different production team than her future husband. Nonetheless, they had known each other their whole lives and were dedicated to this cause, as well as each other, from a young age. While they agreed that their lives and behaviour followed the government recommendations for leading a "low carbon life" (*ditan shenghuo*), they insisted their approach could be more aptly labelled a "peasant way of life" (*nongmin shenghuo*), as it is built on a notion of scarcity rather than plenty.

This self-identification as part of the "peasantry" (*nongmin*) has a long-standing history in China, from Confucian celebrations of these workers as the backbone of the nation to modernist associations that assumed their backwardness, and from the Maoist championing of the peasant classes to the "left behind" (*luohou*) population in contemporary capitalism (see Day 2013, Steinmüller 2011). In defining themselves as "peasants," the same term as in the official state category of rural citizen, they evoked an ideal of an honest, upright and resilient worker while simultaneously implying a more genuine and authentic positionality aligned with citizen ideals throughout Chinese history. While many residents in the countryside play with and even satirize these representations of "ruralism" (Steinmüller 2011), Zhang seemed to inhabit this subject position in a more straightforward manner. Nonetheless, he simultaneously made a living by integrating long-standing local afforestation practices into the accounting infrastructure of digital environmentalism in a way that was not just savvy but, in fact, cutting-edge.

Although they received funding from carbon offset mechanisms, Zhang's family members remained critical of the logic of "carbon sinks" (*tanhui*) to balance urban pollution. When one of their partner organizations – a prominent promoter of corporate carbon projects – came to calculate the payments for the potential future CO_2 absorption of their trees and shrubs and the family's efforts in planting them, the family contested the basis for these calculations, as well as their outcomes in terms of absorption for combatting climate change. Instead, they emphasized their commitment to survival and anti-desertification. Indicating a sign awarded to his family as part of this ongoing carbon sink project, Zhang explained:

> Look, the climate (*qihou*) is not our main priority. First, we look after our bodies and our health, and what the benefits are from improving the environment. Second, we combat the desert, we stop the sand from encroaching and the desert expanding. These are our priorities.

His father, Zhang Senior, described the process of quantification as one in which the carbon partner organization attempted to calculate the future capacity of the plants to absorb carbon dioxide, as well as the workers' clearly delineated work hours – an approach he considered nonsensical considering their family's life-long dedication to afforestation. This concern resulted in extensive negotiations between the family and the organization regarding payment for the carbon afforestation project. Despite eventually resolving this dispute through a compromise of payment for the trees planted, Zhang Senior believed that the calculation criteria should have encompassed not only the effort spent planting and tending to trees in the dunes but also the overall costs of maintaining a life in the desert.

His wife, Zhong, agreed and described an analogy between their approach to planting, growing and raising trees and a mother nurturing her children. In both practices, she maintained that success cannot be guaranteed within the context of harsh environments and therefore projected into uncertain futures. This made projections for carbon absorption nonsensical. Moreover, Zhong felt that the measurements of carbon quantification were misguided as living processes were reduced to a quantified value, as she pointed out that:

> Like a mother feeds her children, we go to water the plants, so what this is worth is impossible to say. And while some children grow big and may be taller than others, a mother cannot be responsible for this. She wants them all to survive. With trees in the desert, this is difficult; the low survival rates are not about neglect.

The unspoken implication of surviving the desert came up over and over in the accounts of this extended family, including in this analogy between seedlings and children, but also in relation to their prioritizing health over the environment. This may well be because the family established the basis for their desert practices through repeated experiences of hardship and loss. However, as the following section shows, the frustration and disappointment the Zhang family and the local community felt with local instantiations of the state, and particularly with the local Forest Bureau, in delivering on immediate green promises further undermined their faith that quantified carbon transfers would give rise to ecological civilization in Wuwei.

Political play with ecological funding

The Zhang family's support for national efforts to develop ecological civilization faltered due to their encounters with the local state. The family's enthusiasm for tree-planting and their conditional cooperation with the local Forestry Bureau resulted in the substantial expansion of their lease to one hundred mu (6.66 hectares) of desert for the next 80 years. Their cooperation with people in what was their former commune production team and the pooling of resources within their hamlet to combat desertification underscored their alignment with the national state's ecological ambitions as a shared endeavour.

However, the Zhangs' interactions with the local Forestry Bureau – and their monetary and carbon calculations – revealed a more ambiguous relationship. The Zhangs claimed that the bureau made empty promises (*kongkou shu baihua*). For instance, the Forestry Bureau had helped finance the watchtower that allowed visiting dignitaries to view the growing green hills emanating from Zhang Senior's home. However, according to the Zhangs, the bureau failed to deliver on the previously negotiated 3000 RMB contribution to construction costs for the actual domestic buildings, leading them to dismiss a TV documentary about their project of "Greening the Desert" that glossed over this conflict as "fake" (*jiade*). A relative expressed their frustration by saying: "The forestry bureau is all promises and deals and photographs and signatures, but then nothing comes. We have to avoid going through them for financing, as they are very annoying (*mafan*) to deal with."

Their mistrust extended to suspicions about a local seedling and tree nursery business engaged in money laundering. Allegedly, this company was "colluding" (*goutong*) with the Forestry Bureau due to having "connections" (*guanxi*), ultimately diverting funds intended for anti-desertification into their own pockets. Another family member explained: "Those higher up don't know. The bureau just takes pictures and pockets the money for illegal (*weifa*) activities. Their kids all go to expensive universities and their families buy urban apartments."

This disenchantment was confined to the lower echelons of the government. The Zhang family only expressed political disappointment with the district government, sparing higher municipal, regional or national tiers from blame. Apportioning responsibility in this way preserves a benevolent image of the central state and simultaneously blames local state actors for experienced injustices, a pattern Guo Xiaolin (2001) has termed the "bifurcated state." This selective trust reflects the enduring wisdom that "Heaven is high and the emperor is far away" (*tiangao huangdiyuan*), thereby containing conflict so it does not reach the upper echelons of the political hierarchy (see also Bruckermann 2019).

The Zhang family viewed the local Forestry Bureau's practices as discouraging small family projects and instead monopolizing financial resources for forestry, casting a shadow over the transparency and fairness of environmental funding. This reveals a tension between upholding the central state's vision and challenging local authorities, where kinship and community-driven environmental work confront opaque financial practices and corruption suspicions, especially within the state forestry farms.

These experiences led the Zhangs to avoid financial dealings with the local Forestry Bureau, as they explicitly preferred engaging directly with urban donors, corporate funders and carbon partner organizations. In the case of carbon calculations, they felt that the calculative mechanisms were opaque and should not be channelled through the Forestry Bureau. Zhang Senior referred explicitly to the negotiation surrounding carbon calculations discussed above, explaining:

> We have a good cooperation with the carbon [partner] organization. They wanted to work with us, because we are not driven by profit. They explained their reasons for calculating the carbon absorption and payments, and we explained why it is not appropriate for the desert. In the end, we reached a solution. This would not have been possible if the Forestry Bureau had stood in between.

The Zhangs thereby extended their local resistance to carbon calculations from the carbon partner organization to the government. As discussed above, the Zhangs had already refused the logic of carbon accounting (i.e., in terms of future carbon absorption) and even wages (i.e., receiving wages for the hours spent afforesting the dunes) as the basis for payment with the carbon partner organization. Instead, they emphasized that remuneration should focus on maintaining human and plant life in the desert through afforestation. This was an expectation they also voiced in relation to the state. Nonetheless, given how opaque the calculative processes for carbon offsets were, they were opposed to entering any carbon-based projects with the local state. The family's experiences of broken promises by the Forestry Bureau and disappointments over payments that never came further undermined the philosophy of promise that ecological civilization and carbon compensation espoused.

Conclusion: carbon values beyond quantification

The emergence of carbon governance in China rests on the state policy goal of establishing an "ecological civilization," a stated aim of the Communist Party for national and global futures. The discourse surrounding ecological civilization can be considered as a form of world-making that unites two unlikely strands of policy: first, the techno-bureaucratic policy of quantification and, second, a re-traditionalizing dynamic fostered by the state that operates through a philosophy of promise. In the last decade, the Wuwei area has been charged with an environmental responsibility that has been scaled up even higher and now encompasses planetary hopes of mitigating climate change through afforestation activities that are tied to financial chains of carbon offsets.

The philosophical and political sources shaping understandings of ecological civilization contribute to shaping the worlds imagined and inhabited by Chinese citizens throughout the country as they encounter environmental events and ecological hotspots. The history of ecological civilization interweaves with philosophical underpinnings of neo-traditionalist readings of religious landscapes and with a neo-socialist formulation of quantifying behaviour and pushing for progress in civilizing campaigns to bring about the coming era.

Yet national ideals do not fully resonate with local strategies. In an area where farmers have been digging ever deeper wells to coax agricultural products from the desiccated soil, the long-term reproduction of life has long been at the forefront of rural concerns. Increasingly, the planting of trees – the creation of China's Green Great Wall – has come to take centre stage in attempts to ward off sandstorms sweeping across North China. In this context, Model Worker Zhang's family of afforestation workers ran a project to green the desert in Northern Gansu, motivated by an ideological melding of red ideology with neo-traditionalist underpinnings.

For the Zhang family and their fellow afforestation workers, their motivations to green the desert melded Maoist ideological positions on labour for the nation with affective commitments to the future of the family and the village, as well as socialist celebrations of their green work as an exemplary contribution to the national project of ecological redress. Moreover, rejecting the logic of quantification and future speculation on which their carbon sequestration payments relied, they instead made the case for environmental work as a matter of survival and reproduction, drawing continuities between their children and their seedlings, which connected social reproduction to their ancestors and their descendants.

Among these afforestation workers, legacies of sacrifice, labour and heroism from the Maoist period of high socialism intermingled with memories of hardship, famine and ecological collapse. Civilizational cosmologies of the proper ordering between humanity and the environment, and of moral expectations for intergenerational conduct and kinship obligations, shaped their visions of the coming era. Concepts of ecological civilization combined the notions of developmental progress with both a philosophical trajectory and socialist obligations based on civil and moral conduct that had taken root in everyday consciousness. In order to be read as exemplary, citizens performed their ecological consciousness by leading low-carbon lives, yet they also resisted this characterization as reductive of their actual subjectivities and hopes regarding a greener future.

In China, carbon logic has become deeply entangled with philosophical, ideological and neo-traditionalist ways of framing future visions, at least in state policy and government discourse. However, the narratives on the frontline, exemplified most explicitly by the Zhang family, tell a different story. Despite conforming to the expectations set by the state for environmental redress, these green workers did not uncritically internalize a carbon-centred worldview (see Whitington 2016). Instead, their actions and perspectives exposed a dissonance with carbon's calculative logic that went beyond the problems of inaccuracy, measurement and deliberation found among their urban, corporate or activist counterparts. Instead of forging a new kind of carbon subjectivity, the gravitational pull of state ideology, anchored in neo-traditionalist and neo-socialist values, exerted a stronger influence than the abstract quantification of carbon metrics on these citizens' self-identification.

In short, grassroots alignment with the state's environmental directives transcended the need for carbon logic among these rural green workers. Even as they generated income from the carbon calculus, they resisted reducing the value of their activities to carbon abstractions and their speculative projections into the future. Instead, afforestation workers in the Chinese countryside, faced with past hardships and future hopes, focussed on sowing the seeds for green transitions despite the risks and uncertainties lying ahead.

Acknowledgements

My heartfelt thanks go to the farmers and foresters in Gansu and their families, who welcomed me into their homes and lives, sharing their experiences and insights on decarbonization campaigns throughout this research. I am also very grateful to Steffen Dalsgaard and the other researchers of the Socio-cultural Carbon project for facilitating stimulating discussions at their workshop in Copenhagen. The exchanges with all participants sparked thought-provoking reflections on the intricacies of carbon and its broader implications. Generous funding for this research was provided by the Max Planck Institute for Social Anthropology, the Department of Social Anthropology at the University of Bergen, the Bergen Foundation and the Department of Social and Cultural Anthropology at the University of Cologne.

Note

1 Identifying personal and place names have been changed, unless specifically requested to be included by the people in question. The names of public figures, policies and campaigns have been maintained in the original form.

References

Bakken, Børge. 2000. *The Exemplary Society: Human Improvement, Social Control, and the Dangers of Modernity in China*. Oxford: Clarendon Press.
Blok, Anders. 2011. "Clash of the Eco-Sciences: Carbon Marketization, Environmental NGOs and Performativity as Politics." *Economy and Society* 40, no. 3 (July): 351–376. https://doi.org/10.1080/03085147.2011.574422

Brandtstädter, Susanne. 2000. "Taking Elias to China (and Leaving Weber at Home): Post-Maoist Transformations and Neo-Traditional Revivals in the Chinese Countryside." *Sociologus* 50, no. 2 (January): 113–143.

Bruckermann, Charlotte. 2019. *Claiming Homes: Confronting Domicide in Rural China*. New York: Berghahn Books.

Bruckermann, Charlotte. 2020. "Green Infrastructure as Financialized Utopia: Carbon Offset Forests in China." In *Uneven Financializations: Connections, Contradictions, Contestations*, edited by Chris Hann and Don Kalb, 86–110. Oxford/New York: Berghahn.

Bruckermann, Charlotte. 2023. "The Pragmatism of Continual Failure: Environmental Policy as Experimentation in China." *JRAI* 29, no. 1 (February): 133–150. https://doi.org/10.1111/1467-9655.13906

Bruckermann, Charlotte. 2024a. "Carbon as Value: Four Short Stories of Ecological Civilization in China." In *Insidious capital: Frontlines of Value at the End of a Global Cycle*, edited by Don Kalb, 96–123. New York/Oxford: Berghahn Books. https://doi.org/10.1515/9781805391579-006

Bruckermann, Charlotte. 2024b. "The Good Life as the Green Life: Digital Environmentalism and Ecological Consciousness in China." *Positions* 32, no. 1 (February): 129–149. https://doi.org/10.1215/10679847-10890010

Callon, Michel. 2009. 'Civilizing Markets: Carbon Trading between in Vitro and in Vivo Experiments'. Accounting, Organizations and Society 34 (3): 535–48. https://doi.org/10.1016/j.aos.2008.04.003

CPC. 2021. http://cpc.people.com.cn/n1/2021/1005/c164113-32245840.html

Dalsgaard, Steffen. 2013 "The Commensurability of Carbon: Making Value and Money of Climate Change." *Hau: Journal of Ethnographic Theory* 3, no. 1 (Spring): 80–98. https://doi.org/10.14318/hau3.1.006

Dalsgaard, Steffen. 2022. "Tales of Carbon Offsets: Between Experiments and Indulgences?" *Journal of Cultural Economy* 15, no. 1 (September): 52–66. https://doi.org/10.1080/17530350.2021.1977675

Day, Alexander F. 2013. *The Peasant in Postsocialist China: History, Politics, and Capitalism*. Cambridge: Cambridge University Press. https://doi.org/10.1017/CBO9781139626309

Ding, Iza. 2020. "Performative Governance." *World Politics* 72, no. 4 (October): 525–556. https://doi.org/10.1017/s0043887120000131

Dynon, Nicholas. 2008. "'Four Civilizations' and the Evolution of Post-Mao Chinese Socialist Ideology." *China Journal* 60 (July): 83–108. https://doi.org/10.1086/tcj.60.20647989

Feuchtwang, Stephan, and Mike Rowlands. 2019. *Civilisation Recast: Theoretical and Historical Perspectives*. Cambridge: Cambridge University Press.

Fraser, Nancy, and Rahel Jaeggi. 2018. *Capitalism: A Conversation in Critical Theory*. 1st edition. Cambridge: Polity.

Graeber, David. 2013. "It is Value That Brings Universes into Being." *Hau: Journal of Ethnographic Theory* 3, no. 2 (Summer): 219–243. https://doi.org/10.14318/hau3.2.012

Guo, Xiaolin. 2001. "Land Expropriation and Rural Conflicts in China." *China Quarterly* 166 (June): 422–439. https://doi.org/10.1017/S0009443901000201

Hansen, Mette Halskov, Li Hongtao, and Rune Svarverud. 2018. "Ecological Civilization: Interpreting the Chinese Past, Projecting the Global Future." *Global Environmental Change* 53 (November): 195–203.

Heilmann, Sebastian and Elizabeth Perry. (ed.). 2011. *Mao's Invisible Hand: The Political Foundations of Adaptive Governance in China*. Cambridge: Cambridge University Press.

Hoffman, Lisa. 2011. "Urban Modeling and Contemporary Technologies of City-Building in China: The Production of Regimes of Green Urbanisms." In *Worlding Cities*, edited by Ananya Roy and Aihwa Ong. Oxford: Blackwell.

Humphrey, Caroline. 1997. "Exemplars and Rules: Aspects the Discourse of Moralities in Mongolia." In *Ethnographies of Moralities*, edited by Signe Howell, 25–48. London: Routledge.
Kalb, Don. 2024. "Introduction." In *Insidious capital: Frontlines of Value at the End of a Global Cycle*, edited by Don Kalb, 1–39. New York/Oxford: Berghahn Books.
Knox, Hannah. 2020. *Thinking Like a Climate: Governing a City in Times of Climate Change*. Durham: Duke University Press.
Liang Guanglin, Zhang Linbo, Li Daiqing, Liu Chengcheng, Luo Shanghua, and Meng Wei. 2017. "Experiences in and Suggestions for Ecological Civilization Construction in Fujian Province." *Strategic Study of Chinese Academy of Engineering* 19, no. 4 (October): 074–078. https://doi.org/10.15302/J-SSCAE-2017.04.012
Lippert, Ingmar. 2018. "On Not Muddling Lunches and Flights: Narrating a Number, Qualculation, and Ontologizing Troubles." *Science and Technology Studies* 31, no. 4 (December): 52–74. https://doi.org/10.23987/sts.66209
Lippert, Ingmar. 2015. "Environment as Datascape: Enacting Emissions Realities in Corporate Carbon Accounting." *Geoforum* 66 (November): 126–135. https://doi.org/10.1016/j.geoforum.2014.09.009
Lippert, Ingmar, Franz Krause, and Niklas Hartmann. 2015. "Environmental Management as Situated Practice." *Geoforum* 66 (November): 107–114. https://doi.org/10.1016/j.geoforum.2015.09.006
Lohman, Larry. 2010. "Uncertainty Markets and Carbon Markets: Variations on Polanyian Themes." *New Political Economy* 15, no. 2 (July): 225–254. https://doi.org/10.1080/13563460903290946
Lora-Wainwright, Anna. 2013. *Fighting for Breath: Living Morally and Dying of Cancer in a Chinese Village*. Honolulu: University of Hawai'i Press.
Lord, Elizabeth. 2020. "Theorizing Socio-environmental Reproduction in China's Countryside and Beyond." *Environment and Planning* 4, no. 4 (November): 1687–1702. https://doi.org/10.1177/2514848620970125
Ma, Tianjie. 2016. "Pan Yue's Vision of Green China." *China Dialogue*, March 8, 2017. https://chinadialogue.net/en/pollution/8695-pan-yue-s-vision-of-green-china/
Moner, Laia. 2021. "Yugong Yishan: Myth, Utopia, and Community in Modern and Contemporary Chinese Art." *China Perspectives* 1 (March): 41–48. https://doi.org/10.4000/chinaperspectives.11398
Neimark, Benjamin, Sango Mahanty, Wolfram Dressler, and Christina Hicks. 2020. "Not *Just* Participation: The Rise of the Eco-Precariat in the Green Economy." *Antipode* 52, no. 2 (January): 496–521. https://doi.org/10.1111/anti.12593
Strathern, Marilyn. 2000. *Audit Cultures*. London: Routledge.
Steinmüller, Hans. 2011. "The Reflective Peephole Method: Ruralism and Awkwardness in the Ethnography of Rural China." *The Australian Journal of Anthropology* 22, no. 2 (August): 220–235. https://doi.org/10.1111/j.1757-6547.2011.00125.x
UNDP. 2021. https://www.undp.org/china/publications/issue-brief-chinas-14th-5-year-plan-spotlighting-climate-environment
Wang, and Li. 2019. http://cpc.people.com.cn/n1/2021/1005/c164113-32245840.html
Wang, Binbin. 2017. *Climate Change in the Chinese Mind: Survey Report 2017*. Beijing: China Center for Climate Change Communication.
Whitington, Jerome. 2016. "Carbon as a Metric of the Human." *PoLAR* 39, no. 1 (May): 46–63. https://doi.org/10.1111/plar.12130
Whitington, Jerome. 2020. "Earth's Date: Climate Change, Thai Carbon Markets, and the Planetary Atmosphere." *American Anthropologist* 122, no. 4 (September): 814–826. https://doi.org/10.1111/aman.13476
Xinhua. 2021. http://www.news.cn/english/2021-10/12/c_1310239983.htm
Zee, Jerry. 2021. *Continent in Dust: Experiments in a Chinese Weather System*. Oakland, CA: University of California Press.

6 The social life of peat carbon and peat frontier making

An ethnographic study of peatland restoration in Central Kalimantan

Anu Lounela

Introduction

One morning in February 2019, it was raining heavily when five men from the restoration evaluation team and I set out by boat on an evaluation trip in the Sei Tobun village.[1] The village is located along the Kahayan River in the southern part of Central Kalimantan province, Indonesia. We were to assess whether the peatland channel-damming project implemented by villagers as part of the restoration of drained peatlands had been properly completed. During our trip, we encountered many problems with the dams, which, according to the restoration experts, were broken or poorly constructed. The villager leading the dam construction questioned the restoration experts' technical knowledge of dam construction, saying that there would be too much flooding if the dams were built so high. Tono, the environmental officer living in the village, became very frustrated and said to the villagers, "Socially and culturally, flooding might be wrong, but from a restoration point of view, flooding is good." The environmental officer's argument that flooding is socially and culturally wrong points to the Ngaju Dayak people's long history of draining wetlands by digging waterways through the peatland. The waterways mark land rights along with crops and trees planted on the drained land. From a restoration point of view (as understood by the experts), it would be right to turn drained peatland back into wetland (flooded with water) by damming. This would prevent fires and store carbon, thus reducing emissions. However, blocking these human-made channels in the wetland would raise the water level, affecting what can grow there, as well as possibly complicating villagers' access to the wetland.

The day after our evaluation trip, we gathered for an evaluation meeting in the climate change mitigation centre, built in the village after 2010. During that meeting, the "stakeholders" (Indonesian Peat Restoration Agency staff, the expert group, regional officials, specific village group representatives, university representatives and donor agency staff) shared their results with the audience, including some of the villagers.

This restoration evaluation was part of the national Peat Restoration Agency (Badan Restorasi Gambut (BRG)) programme that deals with climate change mitigation in Indonesia. As a result of the ongoing and recurrent

DOI: 10.4324/9781003478669-7

peatland fires in Indonesia, and in order to meet national climate change mitigation targets, Indonesian President Joko Widodo (Jokowi) established the BRG in 2016. The BRG was to take the lead in rewetting and restoring drained peatlands, which would reduce forest and peatland fires and mitigate climate change by reducing emissions in seven priority provinces, including Central Kalimantan, where I carried out fieldwork. Due to wetland drainage, deforestation and other factors, Central Kalimantan has experienced wildfire disasters since at least 1973 (Shimamura 2017: 138), but the scale of the fires has escalated since 1997, and now they have become almost a yearly occurrence (ibid. 138; Horton et al. 2021). In light of the fire disasters, the plight of the local people living on disturbed and repeatedly burned peatlands, and the various restoration and conservation projects designed to address the situation, I suggest that it is crucial to explore the negotiations over the boundaries of social orders and socio-material relations in the making of peat into a local carbon store.

This chapter explores the social life of peat carbon in creation and revaluation of peat as a carbon sink and potential carbon commodity through peatland restoration in the context of the National Climate Change (NDC) Targets defined in Indonesian national policy in connection with UN climate change agreements. In this respect, my analysis is based on three concepts that can help us understand how sometimes contrasting imaginaries of carbon are entangled in processes of social and cultural change in Kalimantan.

The first is the concept of a *carbon frontier*, here associated with the making of peat into a carbon store and translating it into potentially valuable a commodity – a process of territorialization that opens up a new type of resource frontier (Vandergeest and Peluso 1995). Rasmussen and Lund have suggested that "frontiers represent [...] the discovery or invention of new resources [...] novel configurations of the relationship between natural resources and institutional orders that happen at particular moments in particular places" (Rasmussen and Lund 2018, 388). Territorialization rearranges space and social orders as strategies of control over resources (ibid.). In a recent article, my colleague Tuomas Tammisto and I proposed that "frontiers are spatial, temporal, and relational *situations* that involve *territorial* processes that qualify landscapes and relations between humans and other beings, such as plants, animals" (Lounela and Tammisto 2021, 5). In this line of thought, territorialization implies processes of maintaining or creating access to landscapes through boundary making involving humans and non-humans (Lounela 2021a). In the case studied here, we see how the villagers use plants and waterways as boundary and territorial markers, pointing to the processual nature of territorialization whereby "boundaries are miniature arenas where people struggle for access to and control of resources and social status" (Sheridan 2016, 31). This is related to the way in which carbon frontiers create new social orders that are not distinct from ecological landscapes – indeed, they often overlap with previous ones and undergo negotiations (Gershon 2019) rather than dissolve them (Rasmussen and Lund 2018, 388).

The second concept draws from work by Amber Huff, who has recently proposed a conceptualization of carbon as a "frictitious commodity" and of what she calls "the repair mode" (2021), in which "[t]he premise that underlies environmental offsetting, carbon forestry and other forms of conservation in the repair mode is that environmental crises arise because nature is a neglected dimension of an 'immanent market-world'" (2021, 3). For example, in my ethnographic case of peat restoration, in terms of the Paris Agreement, peat is framed as a method of carbon storage to mitigate climate change, and peat restoration projects are funded by countries (such as Norway) or transnational agencies (The World Bank and The United Nations).[2]

Third, I am particularly interested in a conceptualization of "the social life of *peat carbon*" and how it unfolds through peat restoration in Central Kalimantan. The idea of the social life of carbon draws from Mahanty et al. (2012), who discuss the creation of "forest carbon" through climate change mitigation and how these processes take on a social life in Southeast Asia. For instance, in the climate change mitigation discourse, peat is translated into an economically valuable carbon commodity aimed at virtual carbon markets. It is meant to be produced through human activities that transform damaged peat into a restored wet landscape. The notion of the social life of carbon suggests looking at the local practices, institutions and labour through which carbon as a natural commodity is translated into social practices. What this translates to in Central Kalimantan is the damming and revegetation of channelled peatlands. In what follows, I focus particularly on the conceptualization of the social life of peat carbon, which importantly links to the other two concepts in supporting an analysis of the role of carbon in transforming the social and cultural life of the Ngaju Dayak living in the Kalimantan peatlands.

In the context of Indonesia and Kalimantan, previous research has to some extent already discussed the confluence of these three concepts, for example, in how the making of the peat restoration involves the production of new subjectivities, such as "forest farmers" (Lounela 2021a). Furthermore, Nancy Peluso has shown how the "smallholder slot" is (re)produced in resource frontiers and becomes a powerful discursive category in Indonesia (Peluso 2017). Also, Anna Tsing has discussed the "salvage frontier" in the context of Kalimantan, which she describes as an imaginary project "capable of molding both places and processes" in which environmental extraction and protection mix (2005, 32–33). Eilenberg introduces the notion of "frontier constellations" whereby "frontier imaginary becomes especially powerful and potent in the regions where resource frontiers and national borders interlock" (Eilenberg 2014, 159–60). Conservation frontier has emerged as a new frontier type in Indonesia (Acciaioli and Sabharwal 2017). In Central Kalimantan, the state, corporations and international funding agencies have brought with them imaginaries of translating nature into money through extraction but also through fixing nature (Castree 2003), as in the case of climate change mitigation schemes (Lounela 2015). This article suggests that the production of the peat carbon frontier through the restoration of peatlands to wetlands operates within the logic of capitalist fixing. This process seeks to address the problems of earlier frontier processes, but it also reorders space and socio-material relations. Thus, ultimately, the peat carbon (frontier) produces new frictions.

I conducted ethnographic fieldwork in Central Kalimantan over seven periods, between 2012 and 2019, in two predominantly Ngaju Dayak villages along the Kapuas and Kahayan rivers, in the city of Palangkaraya, in two provincial capitals, and in Jakarta. This paper is based on fieldwork in the village of Sei Tobun. In this particular village, my fieldwork and research were affected by the 2015 and 2019 wildfires, which burned trees and vegetation, and many residents said they stayed on their land for long periods to protect it from fire, leaving them, as they said, traumatized. These fires prompted many people to work on restoration. My ethnographic fieldwork involved staying in the village and participating in the daily lives of the residents, travelling along the rivers and waterways to gardens, fishing sites and through the disturbed peatlands and participating in climate change mitigation and restoration activities. I also interviewed and spent time with river and channel group leaders and members, village authorities, government officials, climate change mitigation and restoration staff and NGOs at various sites. My research on the social life of peat carbon has revealed overlapping and contested social orders in the process of translating peat into a carbon storage resource. This is partly a result of the messy and "frictitious" nature of this potential commodity in this specific peat carbon frontier, where shifting and overlapping extraction and conservation projects take place.

The Indonesian "Carbon Frontier"

In the UN climate change negotiations, peat wetlands have been identified as potentially playing an important part in mitigating climate change by reducing carbon emissions. Following this development, and because peatland can store large amounts of carbon if restored into wetlands, Indonesia, which has more than 22 million hectares of peatlands, has set the Nationally Determined Contribution (NDC) and implementation strategies (legislation and projects) where peat features heavily (UNDP, 2024). This framing of wetlands as carbon sinks was reinforced by the publication of the Blue Carbon Report in 2009 (United Nations Environment Programme 2009), which focussed on peatlands, marshes and coastal ecology and framed them as valuable "ecosystem services" (Huff 2021).

Despite this potential, when degraded, peatland becomes a source of carbon emissions. Peat is formed in a long process of decomposition of dead plant material, waterlogged and accumulated in peat bogs. These can be tens of metres deep, and in some contexts, the bogs form a peat swamp forest, such as is the case in riverine Central Kalimantan (Shimamura 2017, 123–124). Peat swamp forests in the tropics are mostly inundated with water and store carbon in that state. But when peatland is drained, it releases carbon that has accumulated over a long period of time. It then also burns easily, releasing huge amounts of carbon dioxide into the atmosphere along with smoke and haze (Ibid., 137–138). For these reasons, restored peatland can potentially become a natural commodity in carbon markets. As UNOPS (The United Nations Office for Project Services) describes the Indonesian peatland restoration: "Over time, efforts like this across the country will simultaneously decrease destructive peatland fires and protect our planet by leading to measurable reductions in greenhouse gas emissions" (UNOPS, 2024).

Indonesia has one of the largest tropical wetland areas in the world, with peatlands covering about 22.5 million hectares and wetlands about 4.2 million hectares (CIFOR 2021). Central Kalimantan is one of five provinces on the island of Borneo. Its environmental situation has deteriorated rapidly since the New Order government (1967–1998) due to logging, the construction of agricultural canal infrastructures and the rapid expansion of oil palm plantations. The peatlands of Central Kalimantan cover an area of about 2.6 million hectares. In Pulang Pisau regency – the site of this study – 67% of the 983,000 hectares of land is peatland, much of it degraded. One report states that between 2003 and 2019, around one million hectares were burned in the district, and in 2009, 2015 and 2019, around 100,000 hectares were burned, with some areas burning repeatedly (Hapsari et al. 2020, 8). The area has a long history of wetland drainage, channelling and forest degradation (Lounela 2021b; Nygren and Lounela 2023), resulting in regular wildfires (see Galudra et al. 2011; Horton et al. 2021; Lounela 2021b). In response, peat restoration and climate change mitigation projects have proliferated in the area in recent years (MMCKalteng 2024a, 2024b).

Since mid-2000, Indonesian peatlands, about half of which are degraded, have become the second most important restoration category in Indonesia's forest restoration and rehabilitation policy (Fisher et al. 2023). Following the 2007 UN Climate Change Conference in Bali (COP13), Indonesia began negotiations with foreign donors and enacted regulations to pave the way for climate change mitigation pilot projects in the country. It also committed to significant greenhouse gas emission reduction targets. In 2010, Indonesia signed a US$1 billion agreement with Norway to facilitate emission reductions through various programmes. In addition to the agreement with Norway, Indonesia was opened to a large number of REDD+ (Reducing Emissions from Deforestation and Forest Degradation) "demonstration activities," pilot projects to mitigate climate change. These experiments on the impact of carbon trading schemes in forests were meant to help develop REDD+ mechanisms. As a result, Indonesia has been the site of tens of REDD+ pilot projects (Resosudarmo et al. 2014). In 2010, Central Kalimantan was designated as a pilot climate change province by the President of Indonesia, prioritizing climate change mitigation through REDD+ projects. Although some of the pilot projects were halted due to various problems (see Lounela 2015), some projects are now selling carbon credits (Fisher et al. 2023).

As mentioned at the beginning of this article, the National Peat Restoration Agency (BRG) was established in 2016 and began peat restoration activities in 2017. This government programme is linked to Indonesia's National

Development Plan regulation, which outlines national climate change mitigation (26% emission reductions by 2019), with forestry and peat being one of the five priority areas (PP-Peraturan Presiden Republik Indonesia No 2 Tahun 2015, 173). The BRG has received a new mandate for peatland restoration for the period 2020–2024, although the targets of the previous period were not reached (Jong 2021). Some Indonesian scholars and BRG officials argue that funding for peat restoration is considered to be too low (Sari et al. 2021). Much of it comes from international donors, with Indonesia providing part of the funding from the national budget. Therefore, carbon markets are envisaged as a future funding potential (Sari et al. 2021).

International donors, such as USAID and UN-affiliated agencies, see climate change mitigation (such as REDD+) as a "social project" that is both about environmental conservation and benefiting local communities by providing them with economic benefits, thus commodifying both nature and social relations (Howell 2014, 270). There are various ways to bring economic benefits to local people, such as planting rubber trees and forming rubber cooperatives, connecting them with private corporations and, more recently, bringing social forestry programmes to local communities to restore peat and replant peatlands with timber. The idea is that the plantations will bring economic benefits to local people – that is, by helping people grow "valuable commodities" (Lounela 2015).

In 2016, the BRG began working with the USAID programme LESTARI (2015–2020), operating in the Katingan-Kahayan area (USAID 2020a). The BRG programme included policy and legislative development, media work and training, and activities related to canal blocking, fire mitigation, revegetation and rewetting. As such, USAID mobilized resources for, among others, the BRG's rehabilitation programme of canal and channel blocking:

> In support of this [BRG] initiative, LESTARI facilitated a model stakeholder engagement activity involving the facilitation of FPIC[3] in villages covering around 30,000 ha of degraded peatland. This peatland is part of an area that covers less than 5% of the province yet accounted for 30% of all fire impacts in 2015. FPIC facilitation ensured that communities are well informed about canal blocking, have an opportunity to provide inputs and give their willing consent to construct, maintain and protect the dams. Notably, local communities were able to influence the design of dams so that their small boats could pass through spillways in order to maintain their livelihoods. LESTARI provided technical and financial support for the FPIC process mediated by the district-level multi stakeholder forum named Forum Hapakat Lestari. It adhered to both USAID and BRG social safeguard guidelines for FPIC.
> (USAID 2020a, 50)

The canal and channel blocking activities aim to rewet the drained peatland to make it a carbon store, thereby mitigating both climate change and peat fires. The programme responds to Indonesia's need to set national climate change mitigation targets under the Paris Agreement and to potentially participate in voluntary carbon trading. However, translating peat carbon into local social activities (blocking channels, revegetation, etc.) means that local people work to restore drained peatlands, which USAID and the BRG assume, takes local people's perspectives into account (FPIC). I explore this in more detail below.

As a new type of resource, peat carbon contributes to the making of new commodity frontiers (Rasmussen and Lund 2018). However, here, peat is situated within a long history of violent natural resource making, such as extensive timber logging (Tsing 2005) and the environmentally destructive state agricultural programme the Mega Rice Project (MRP) (see below on Wetland as a property frontier: emergence of *gambut* (peat)), which in Central Kalimantan contributed to violent frontier making that involves blurring the boundaries between legal and illegal (McCarthy 2013).

In the following section, I conceptualize the nature of peat carbon and how this influences the making of the carbon frontier and the social life it engenders.

Theory: the making of the peat carbon frontier and the social life of peat carbon

To appreciate the emergence of a peat carbon frontier and how peat carbon takes on a social life in Central Kalimantan, it is necessary to revisit a range of debates about territorialization in the region. As Nancy Peluso has argued, it is crucial to study the present commodification of nature because "some institution or configuration of historical forces needs to bring them [natures] into marketable being, and the new relations created in the processes are critical" (2012, 85). Yet, how relations are formed and maintained in the production of the peat carbon frontier leads to a discussion of how carbon takes on a social life. Rather than exploring the measurement and calculation of a carbon commodity aimed at a carbon market exchange, I draw on key regional and anthropological literature to explore the less studied social practices (relationships, labour and valuations) of peat carbon production on the ground. This approach provides tools for analysing peatland restoration involving villagers, peat and restoration experts, state officials, NGOs and university scholars.

As mentioned above, the most common discourse employed in rendering peat a carbon commodity is based on the idea that peatland territories, when restored to wetlands, are globally important for carbon storage and have a high economic value that can be calculated (in terms of the amount of carbon) and translated into money, benefiting nature, climate change mitigation plans and local populations (Sari et al. 2021). Larry Lohmann has explored the economic calculation of carbon (carbon accounting and cost–benefit analysis) as a technique through the metaphor of "framing," showing how, as techniques, they engage with "multitudes of new spaces, subjects and objects in their work to 'make things the same'" (2009, 529), requiring the expert knowledge while at the same time generating "resistance." These processes subject peat landscapes to new forms of governance and authority, reproduce social relations and require language work and knowledge production. For example, institutions, including state agencies such as the BRG and collaborating donors (e.g. USAID and the UN), organize legislative, media and policy work at different scales, and they organize socialization events and train local people to obtain knowledge on climate change mitigation and environmental issues.

Knowledge and technical practices play an important part in territorialization through restoration, when dams are built and channels are blocked to raise the water level in the peatland. In addition to the knowledge and practices involved in building the dams, these also extend to the choice of specific plants to be grown on the peatland – the revegetation part of the restoration. The qualities of the plants (trees or other vegetation) have a huge impact on the lives of the local people, but also on the more-than-human entanglements in the peat landscape. This became evident during my fieldwork when fast-growing sengon (*Falcataria moluccana*) timber became part of the restoration process.

Recently, Besky and Padwe (2016, 9) have discussed how territory-making processes are entangled with plants. In this line of thought, materialities such as peat, water and plants can be understood as agents that become part of the formation of social relations through boundary-making in the making of territory (Brighenti 2010; Sheridan 2016). Territory, thus, refers to the processes by which social relations and spatial boundaries are formed, the processes of marking that express access (or inaccessibility), and the ways in which things and beings, such as plants, become part of a territory in the processes. This definition of territory critically challenges earlier scholarly discussions that defined it as a fixed material space and object (Elden 2010) and points to the multi-species relations that are formed in the processes (Tsing 2015). My point here is that the categorization and translation of peatlands into carbon storage territories transform multi-species relations and socialities in unpredictable and sometimes contentious ways, which are critical to both restoration and local ways of life.

As mentioned above, Amber Huff (2021) has proposed understanding carbon as a partially "frictitious commodity." The inspiration for this proposition originates from Karl Polanyi (2001, 76), who suggested that when land, labour and money – things previously not considered commodities – became part of market exchange, they became fictitious commodities. During this process, they were removed from the rest of society and social relations – that is, they became socially disembedded from socio-material relations. Amber Huff argues that this discussion needs to be updated, and she suggests that carbon is only partially a fictitious commodity. Carbon is fictitious in the sense that it becomes a commodity on the market, where it previously held no monetary value. However, Huff argues that carbon offsetting is simultaneously a "fictitious" (not previously monetized) and "frictitious" commodity – a transient commodity concretized through plans and projects and in different forms of nature. This, I assume, means that once concretized and materialized through plans and projects (such as through peatland restoration), it also becomes a situated commodity (Peluso 2012) in the sense that it begins to have a social life in a specific context (i.e. place) and new frictions emerge (Tsing 2005). For example, Mahanty et al. (2012), discussing the social life of forest carbon, convincingly state that:

> Transforming forest carbon into a commodity requires its privatisation, its individuation or delineation into individual units for trade, its valuation in monetary terms, and finally its displacement from the labour and resource context in which it was 'produced' (Castree 2003). The resulting commodity chain for forest carbon, which appears as an intangible unit for exchange, ties its production back to the material realm (forests, land and labour), whose benefits and risks are distributed through socio-political and economic relations and institutions, from the local to the international level (Ribot 1998).
>
> (Mahanty et al. 2012, 661)

Similarly, the "social life of peat carbon" refers to the ways in which peat carbon is produced as local residents work on or live with peatlands and participate in their restoration, with the potential risks and changes that result from restoration activities. To take the example from my field site mentioned in the introduction, people block canals to raise water levels and stop drainage, but these activities bring along new risks and burdens in terms of long-term tenure systems and access to land and resources.

Resource frontiers have been seen as representations of the dissolution of existing social orders, where landscapes are imagined and represented as "empty" or underused, while territorialization is thought to constitute new social orders (Rasmussen and Lund 2018). However, in this chapter, I put forth the idea that the making of peat carbon and the peat carbon frontier contribute to multiple social orders and value regimes, which overlap and are in tension with one another, and their boundaries are constantly negotiated.

Rawa: flexibility and mobility in the peat wetlandscape

Sei Tobun, a village of about 2,800 inhabitants, of whom a majority are indigenous Ngaju Dayak, is located along the banks of the Kahayan River in the southern part of Central Kalimantan province on the Indonesian side of Borneo. Ngaju literally means "upstream" and is an exonym for various groups who identify themselves as river basin groups and who speak the Ngaju language (Schiller 1997, 186). The central settlement of the village is situated along one of the largest rivers – the Kahayan – in Central Kalimantan, made of wooden plank houses built atop wooden pillars facing both the river banks and the road. My fieldwork focussed on what is considered the centre of the village, consisting of five neighbourhoods (*rukun tetangga*) located on the bank of the Kahayan River, while the entire village consists of a total of 11 neighbourhoods.

The Ngaju were originally horticulturalists and shifting cultivators. Because they fished, hunted and gathered forest products, they were relatively mobile, sometimes living in forest areas (*kaleka*), fishing in swamp areas, making small ponds in the forests (*beje*) and preserving some forest areas from human use. The Ngaju believe certain forests are sacred and that spirits live in some of the trees, stones, fields and so forth in the landscapes. Gathering forest products, making small fishponds and sharing the harvest with people living nearby were important components in creating socio-material landscapes characterized by mobility and flexibility. These subsistence practices and mobility, and the way they settled for different periods of time in longhouses, in individual houses along the rivers or in swidden fields and gardens, were important in creating the Ngaju Dayak moral values of flexibility, autonomy, sharing and reciprocity (Lounela 2019, 2021b).

Here, the wetland is locally known as *rawa*. In this specific wetland, peat and water are in a constant state of flux. An old man told me that they used a small wooden boat, which they stood in and pushed with a pole, to move further into the swamp forest to collect forest products. In general, people found it difficult to control the water flows in the *rawa*. However, it is crucial for the villagers to know how water moves in the Kahayan River and in the small rivers that flow from the Kahayan through the peatlands, because it is the movement of the water that allows them to access and maintain their gardens and fishing grounds (Lounela 2021b). Living with the *rawa* has given them an intimate knowledge of how to navigate their lives in this wet landscape, but shifting and overlapping frontierization has transformed it into a highly unpredictable one.

Wetland as a property frontier: emergence of gambut (peat)

During the Dutch colonial period, German missionaries and colonial rulers urged the Ngaju Dayak to move from smaller rivers and forest areas and settle along the Kahayan River. Longhouses were constructed, and some large families settled in them. Pak Erlin, who is about 80 years old, explained to me that it was the Dutch and the missionaries who pushed the Ngaju Dayak to settle along the Kahayan River instead of living in scattered houses along the smaller tributaries:

> My grandfather was the head/chief [mengepelai] there [on the river], it was not like staying in the house, but they were ordered to open land on the rivers, small rivers, it was wetland (rawa), the river went through a wetland. Basically, the missionaries ordered that those Dayak people who stayed on the rivers should go back to the settlement [kampung].

In the early twentieth century, the Ngaju Dayak expanded small rivers into longer waterways deeper into the wetlands for various reasons. This was most probably partly due to Dutch policy, which allowed outsiders to extract rattan from Dayak territories and began promoting individual property rights in the mid-nineteenth century to encourage resource extraction. The Dutch granted personal certificates recognizing ownership of rattan forests to those who had opened access to the forest by building roads or digging waterways (Knapen 2001, 362). The Dutch also instituted land laws that would grant private land rights by introducing the Agrarian Law of 1875 (*Agrarische wet 1875*). After the official inauguration of the Domain Declaration (*Domeinverklaring*) in 1875, the Dutch colony became the legally recognized holder of rights to all wasteland and forests unless the Dayak (or others) could prove ownership (Ibid., 362). Even today, people can claim rights to land with *Domeinverklaring* certificates. By the early twentieth century, both Banjar and Dayak kin groups held so-called *tatah* and *sungai* (channel and river) rights (Vergouwen 1921). The demarcation of much of the land in Kalimantan as "wasteland" was used to mark private land rights by digging waterways; the remaining land became state land under the new law. This left many local communities without formal land rights, a situation that continues to this day (Kelly and Peluso 2015).

During my fieldwork, many villagers – especially the river heads or respected elders – told me that the men who first worked on the rivers were considered the "owners" of the rivers, who then distributed land to their family members. Thus, the groom's family was obliged to give the wife's family a piece of land when they married, land often situated along a river. Many rivers had two owners on different banks of the river who were not necessarily related. Two families would work together to extend the river and distribute plots on either side of it, thus inscribing genealogies to the river banks (Lounela 2021b).

The Ngaju Dayak became accustomed to extending rivers deeper through peat wetlands, draining them for cultivation, and planting rice, then cassava and finally rubber along the waterways. Tenure rights to peatlands were marked with dykes, waterways and rubber trees, the latter indicating long-term rights as is also common elsewhere in Borneo (Dove 2011). Behind the "private" rights, however, we can find bundles of rights (Lounela 2009). For example, large families could collectively harvest fruit trees through various arrangements, illustrating the flexibility of these arrangements. The Ngaju Dayak formed kinship and other social relationships along the cultivated rivers, indexing the matrix of social relationships through the landscape. These social relationships are intertwined with the fluctuations and intermingling of water and land (Lounela 2021b).

A peat wetland is a specific assemblage of materials. Before my fieldwork, I could not imagine anyone digging rivers by hand through the wetland. In 2016, I participated in the damming of a channel with a group of men from the village. We travelled by boat about nine kilometres from the settlement into the wetland forest, which had just lost its tree cover in the 2015 forest fires. The men were cutting small galam trees to build a dam across the canal. One of the men took a kind of machete (*parang*), cut the peat soil and lifted it into a sack – the soil was light. Then they filled plastic bags with peat and placed them between wooden fences in the river. When I saw that the peat was so light that it could easily be cut into pieces, placed in the sacks and lifted off the ground, the stories of opening up rivers and widening them by hand began to make sense to me.

This reterritorialization of the wetlands took a new turn in the 1990s. At that time, President Suharto launched the MRP, a 1.4 million-hectare rice cultivation scheme. Within two years, 4,000 square kilometres of canal infrastructure had been built, cutting through the wetlands between the main rivers (the Kapuas and Barito Rivers) in the Central Kalimantan province, causing enormous environmental damage (McCarthy 2013). This agricultural frontier extended into the village, and a large canal was built through the settlement and the wetlands. Some villagers resisted the MRP because they had to give up land along the canal with minimal compensation. The MRP officially came to an end in 1999, after logging had depleted large areas of forest and rice harvests had failed due to the acidic and poor peat soils. Also, President Suharto's presidency came to an abrupt end. In 1997–1998, large-scale forest fires broke out in the area, effectively putting an end to the project (Galudra et al. 2011, 434). Since then, wildfires have occurred regularly, as the drained peatland is highly flammable and fragile during the dry season (Horton et al. 2021).

In 2005, Sei Tobun village received support from the regional government (the governor was a Ngaju man with ties to the village) to use excavators to lengthen and deepen the rivers in order to expand their rubber gardens. In this way, the villagers secured land for subsequent generations through land distribution along the rivers and prevented oil palm expansion into their villages. New people (even from outside the village) were integrated into what was increasingly called channel groups (rather than river groups). The channels served as important waterways to distant locations on the peatland, residents' gardening and fishing sites, and they also marked land rights, which aimed to ensure long-term relations with the landscape by manifesting their rights to land along the channels.

Today, villagers keep the channels "clean" (of vegetation) and open, either with the aid of excavators or by hand. This work aims to stabilize their access to distant gardens and fishing or foraging sites. In terms of infrastructure, the channels form a network that helps villagers connect to these sites as well as to each other and to relevant plants. Gardens are places where social relations are enacted, where couples and extended families can work the gardens together, tapping rubber or collecting rattan and forest products, thus reproducing socio-material relations. As I have noted elsewhere, the Ngaju Dayak people highly value relationality and flexibility in their relationships and living in the watery environment (Lounela 2021b). In this way, they have been able to continue their relatively autonomous way of life, deciding their own labour time in the rubber gardens or fishing and collecting forest products, combining subsistence and market-based livelihoods.

The social life of peat carbon: contestations and frictions

In this section, I explore the social life of peat carbon within the carbon frontier in Sei Tobun village, where climate change mitigation activities began in 2010, with development agencies working to implement REDD+ and climate change mitigation projects. Indonesia's first climate communication centre (called "bamboo house" by the villagers) was launched in 2012 in Sei Tobun. This centre was the result of work carried out by UNOPS under the auspices of UNESCO.

USAID began supporting projects in the area in 2010. Initially, USAID supported two different climate change mitigation projects in Sei Tobun (USAID-IFACS, Indonesia Forest and Climate Support 2010–2015). The first of these was the development of a rubber economy, which consisted – not surprisingly – of planting rubber trees (*Hevea brasiliensis*) and producing latex, but also the building of a latex warehouse, establishing a rubber cooperative in the village and promoting cooperation with banks and companies.

The second was the legal establishment of the Village Forest area (*hutan desa*) – an area of approximately 7,025 hectares (almost half of the village area), granted by the Ministry of Forestry. It is an area managed by the village on state forest land, as regulated by the Ministry of Forestry in 2008 legislation. In Sei Tobun, the forest land was categorized by the state as a "production forest." However, during the "logging season," as the villagers called it, the forests were largely cut down when timber companies entered the area, especially in the 1970s and 1980s. The area was "cleared" of the rest of its trees during the MRP mentioned earlier (Jewitt et al. 2014). The state category of "production forest" (*hutan produksi*) means that the national government has the right to allocate concession areas for logging companies. As this area is peatland, it was later recategorized as a protected forest (*hutan lindung*) by the state. Working with the Central Kalimantan-based NGO, the villagers mapped the area and applied for Village Forest status, which was granted by the government in 2012. The villagers had to form a Village Forest Management Organization (LPHD – Lembaga Pengelolaan Hutan Desa), which would be responsible for managing and protecting the Village Forest area and for ensuring that only non-timber activities are conducted in the area. USAID funded this work through the local NGO, which worked with the village head and villagers at the time (Lounela 2019).

When Joko Widodo became president in 2014, he sought to make national peatland restoration a priority. After 2017, the BRG mobilized and supported the blocking of channels in the village with wooden dams, the formation of village-based firefighting teams, the construction of wells that could be used to mitigate fires and some livelihood programmes such as fishponds in collaboration with USAID's second programme in the region, LESTARI (USAID 2020a). The restoration programme included three components (3R): revegetation, rewetting and revitalization.

Thinking about the social life of carbon, revegetation is a particularly interesting part of the restoration. In 2016, Joko Widodo inaugurated the plywood factory on the opposite side of the river from the settlement. Villagers reported that the president strongly encouraged people to plant the fast-growing timber species sengon on the state production forest land in order to feed the plywood factory. Thus, villagers started planting sengon seedlings after a new forest peasant organization was formed and granted funds to do the work, while some people did so on their own initiative. The planting area was partly the same as the area where the dams were built on the channels (Lounela and Wilenius 2023).

In February 2019, I participated in the evaluation of peatland restoration activities supported by the BRG in Sei Tobun, as mentioned at the beginning of this article. The trip was organized and led by Team Nine – a special team under the district's multistakeholder forum, Hapakat Lestari, which facilitated the process (USAID 2020b, 10). I joined a five-member team consisting of two men from the village, two provincial experts (Team Nine) and Tono from the Department of the Environment (KLH), who lived in the village's climate change information centre.

On that day, I met the men at the climate change information centre and we drove to the mouth of the river on a paved road that runs through the settlement. At the mouth of the river, we met Banum, a villager from the Banjar region (in the southern part of Kalimantan). Banum led the damming of this particular river with the local river group for the BRG project. The BRG funded the work. He took us by boat along the *handil*, a river that villagers had widened into a long channel with the aid of excavators about 20 years ago. Our goal was to inspect and assess five wooden dams that had been built to block the flow of water into the Kahayan River.

On our way, it became clear that several of the dams were either broken or poorly constructed. Tono took the most active role in the evaluation and seemed particularly frustrated when he discovered that the wooden dam structures, designed to be erected 40 centimetres below the surface of the land, were actually deeper or broken. As a result, water appeared to flow freely over the first dam. Tono explained that the dam was not watertight because not enough earth had been used in its construction. The river owner, Poli, told me that people using boats had broken the dam by going over it at high speeds.

The spillway of the next dam was also broken and too low. Banum explained that the dam's spillway had broken because of the heavy flow of water from upstream. I found this strange because, according to the river owner, this channel ended in the middle of the peat swamp and did not connect to other waterways further into the wetland. I knew that there was a waterway that cut horizontally across some of these rivers and connected them to a large canal. Tono and Banum had a heated argument. Banum said in an angry voice that if the spillway was 40 centimetres below the land's surface, the area would be flooded. Tono replied that this was the purpose of the dam and that any flooding would be a good thing. He added: "We don't want the peatlands to be burned, and if the peatlands are wet, they will restore themselves."

We continued our journey and found oil in the water at the next dam. This dam was also broken. Together, the men explained to us that illegal loggers had been transporting logs from Sebangau, a nature reserve situated upstream. It was now clear that the channel extended at least six kilometres from the settlement and connected to a horizontal waterway bordering a state forest, which would probably explain at least part of the large volume of water.

146 *The Cultural Complexity of Carbon*

This horizontal waterway was connected to a large canal built in the mid-1990s as part of the MRP. This opened the way to the Sebangau area, from where illegal loggers dumped timber into the Kahayan River.

"This is a disaster," said the river owner. "We should report it to the police," said Aris from Team Nine. "No, no. They are supporting it here," argued his colleague, Tanam. "We should report it directly to Jakarta," said Toto. After a brief discussion, they decided to return to the village. The men from Team Nine would write a report, and we would all gather for an evaluation meeting with other river owners and stakeholders. The meeting will be held in the climate change information centre in the village.

Tono's assertion that I opened this chapter with – that flooding is socio-culturally wrong but right in terms of restoration – illustrates how the experts and the villagers had different perceptions about the purpose of the rewetting, or how it should be done, and the consequences of the dam building. As mentioned earlier, the villagers regarded rivers and channels as important waterways because they needed access to distant peatlands for cultivation, collection of forest products, fishing and hunting. Banum's argument revealed two concerns: first, about the qualities of the plants that should be planted on the peat wetland that would be inundated for most of the time; second, about access to distant locations if the dams were difficult to pass in wooden boats (even though the BRG had designed the dams to allow passage) (Figure 6.1).

Figure 6.1 Villager passing the dam on a boat during the restoration assessment in February 2019.

Source: Photograph by author.

Damming was, therefore, a social as well as a material, economic and cultural issue: creating a peat carbon commodity through the damming of channels contested the villagers' historically formed practices on the riverine peatland. Thus, while villagers were damming the channels as part of the BRG restoration programme, they often also opened up and cleaned them. Such shifting territorialization – opening and blocking the waterways back and forth – shows how unstable peat carbon is as a commodity form. This became even more apparent to me in 2019 when wildfires re-emerged in the area, and the peatland once again became an "emitter." As such, efforts to stabilize and territorialize carbon commodities through restoration remain fragile.

A further complication appeared during the rainy season in 2019, when the water levels rose and it seemed that the sengon trees on the peatland could not withstand the floods. They turned yellow and distorted, and the villagers feared that they might be difficult to sell to the factory. In September 2019, eight months after the dam-building evaluation trip and the event described above, I returned once again to the village. Now, many of the sengon trees had been burned by fires, and the villagers were surprised that the sengon species had died so easily. It was in the middle of the "fire season," as the villagers call it because of the regularity of the wildfires, and yellowish smoke was visible above the settlement. This time the fire destroyed most of the villagers' sengon trees planted after the inauguration of the plywood factory 4 years ago, some old rubber gardens and parts of the dams. As many villagers told me, the prolonged fires (which hide for a long time under the surface of the peat) had left the villagers traumatized. This was due to their experiences of working in the fields in the heat and toxic smoke to try to prevent fires in the shallow peat soil, as well as witnessing the loss of many trees essential to their livelihoods. Following these events, the villagers wanted to build dams to prevent fires. Peat restoration in terms of territorialization was thus linked to a combination of the qualities of plant materialities, weather and climate patterns, and human planning of landscape arrangements stemming from local, national and international agendas related to the valuation of (peat) carbon (Figure 6.2).

Conclusion: the social life of peat carbon and the making of new value regimes in the carbon frontier

In this chapter, I discussed the production of peat carbon in a specific situation of peat carbon frontier making – a new type of resource frontier. I focussed on the social life associated with turning this peatland into a carbon sink, especially as it unfolds in the restoration of a degraded and drained peatland. As noted earlier, this ethnography of peat restoration, which captures the social life of carbon in wetlands, demonstrates the importance of studying the potential commodification of nature and the changes experienced in peatlands as they shift from being considered sources of carbon emissions or carbon storage, the frictitious nature of this potential commodity and the impact peat carbon has on the lives of local residents.

Peat carbon frontier-making is connected to the global climate change agenda, national climate change politics, potential carbon markets and

148 *The Cultural Complexity of Carbon*

Figure 6.2 Travelling along the newly cleaned channel in the village after the wildfires in the fall of 2019.

Source: Photograph by author.

emissions compensation. Here, peat carbon is potentially a virtual commodity, but it is situated in processes of territorialization and reterritorialization, which makes it unstable. (Re)Territorialization through peat restoration reorders space by closing rivers, with the expectation of rising water levels on peatland, affecting future tenure and planting systems and access to distant locations. Territorialization also reorders space through revegetation, which introduces plants (sengon) along with new social institutions (forest peasant groups) and social orders. Rather than dissolving previous social orders (Rasmussen and Lund 2018), peat carbon frontier-making produces new socio-material configurations that overlap with previous ones, creating situations where people are constantly negotiating the boundaries of these social orders (Gershon 2019), as well as the boundaries of the territories they co-inhabit with plants and animals (Lounela 2021a; Sheridan 2016). While peat carbon is potentially virtual, it has a social life and is literally located in a specific socio-material place and situation (Dalsgaard 2013, 2016).

In socio-material terms, peat carbon has specific properties that affect the formation of the peat carbon frontier and how its social life unfolds. Recent decades have shown how flammable drained peatlands are and how vulnerable some plants are in this environment. Raising the water level and returning the peatland to the wetland means that some plant species do well in this peat environment, while others do not. In this socio-ecological system, peat carbon proves to be highly unstable and fragile.

The social life of peat carbon as a concept unfolds the social and unintended consequences of this restoration infrastructure. In this regard, peat carbon is only partially a fictitious commodity because of its intangible nature, and it could be understood as a frictitious commodity since it assumes a concrete form and comes alive through concrete plans and projects (Huff 2021, see also Dalsgaard 2013). As mentioned earlier, territorialization operates through knowledge creation as new forms of knowledge are introduced through restoration – for example, with the blocking of channels, for which specific forms of knowledge and technical practices must be adopted. Team Nine members, state officials, climate change mitigation agents, corporate actors and scientists form a "translational zone" (Neale 2022) that socializes people about the "right" ways to dam the channels in order to stop fires and mitigate climate change. This knowledge became contested on the basis of the valuations and knowledge that local people had generated on the basis of their long history of living on the peat wetlands, which should provide them with a future basis to continue living there.

Ideally, each successfully blocked channel in the peatland results in carbon being stored in the peat, which can then be measured in terms of how much carbon is stored in the peatland. This can in turn be counted as climate change mitigation, either against national emission reduction targets or in global carbon markets. I have not discussed how this measurement is made and how restored peatlands match the monetary value of peat carbon in the carbon market. However, I have shown how the social life of peat carbon – the creation of peat carbon through restoration on the ground – is intertwined with the lives of people who have long and intimate relationships with the surrounding peatland and who value their flexible and autonomous socio-material relations with the watery peat landscape and its other inhabitants. Restoration may therefore contradict people's valuations of the socio-material relations and forms that are historically made in a particular context. The challenges that peat restoration poses to local tenure systems and human-plant-water relationships therefore have profound implications. The social life of peat carbon is fraught with frictions due to overlapping and multi-scalar social orders, multi-species relations and forms of governance that are entangled with natural (fire) disasters.

In sum, I have discussed the making of the peat carbon frontier in Central Kalimantan in terms of the social life of peat carbon as it plays out in the restoration of degraded peatlands. The case illustrates how different value regimes, social orders and social relations, including humans and nonhumans, are in tension and continually negotiated in the making of the peat carbon frontier. The commodification of nature, such as through the production of peat carbon and its social life, and how this relates to the production of territory and the carbon frontier, is a critical issue in the face of environmental change and the creation of climate governance regimes.

Acknowledgements

I would like to thank the editors for inviting me to contribute to this book and for their very valuable comments. I would also like to thank the participants in the book workshop for their insightful comments. In particular, I would like to thank Steffen Dalsgaard. I like to thank Mira Käkönen, my project colleague, for her important insights. I also thank Sango Mahanty for her valuable comments. In addition, I want to thank my colleagues at the University of Jyväskylä for their comments on my presentation. However, the responsibility for the content of the text lies with me. Finally, I would like to thank the Kone Foundation for funding the "Repair and responsibility in ruined environments" (2024–2027) and "New regimes of commodification and state formation on the resource frontier of Southeast Asia" (2018–2023) projects, both of which contributed to this research.

Notes

1. The name of the village and the interlocutors are pseudonyms to protect the privacy of the people.
2. The Norway-Indonesia deal shows how the repair mode works through private agreements between the nations, such as Indonesia and Norway (Regjeringen, 2024).
3. Free, Prior and Informed Consent.

References

Acciaioli, Greg, and Akal Sabharwal. 2017. "Frontierization and Defrontierization: Reconceptualizing Frontier Frames in Indonesia and India." In *Transnational Frontiers of Asia and Latin America since 1800*, edited by Jaime Tejada and Bradley Tatar, 431–436. London: Routledge,.

Besky, Sarah, and Jonathan Padwe. 2016. "Placing Plants in Territory." *Environment and Society* 7 (1): 9–28. https://doi.org/10.3167/ares.2016.070102

Brighenti, Andrea Mubi. 2010. "On Territorology: Towards a General Science of Territory." *Theory, Culture & Society* 27 (1): 52–72. https://doi.org/10.1177/0263276409350357

Castree, Noel. 2003. "Commodifying What Nature?" *Progress in Human Geography* 27 (3): 273–297. https://doi.org/10.1191/0309132503ph428oa

CIFOR (Center for International Forest Research). 2021. *Global wetlands*. Last Accessed August 13, 2024. https://www2.cifor.org/global-wetlands

Dalsgaard, Steffen. 2013. "The Commensurability of Carbon: Making Value and Money of Climate Change." *HAU journal of ethnographic theory* 3 (1): 80–98. https://doi.org/10.14318/hau3.1.006

Dalsgaard, Steffen. 2016. "Carbon Valuation: Alternatives, Alternations and Lateral Measures?" *Valuation Studies* 4 (1): 67–91. https://doi.org/10.3384/VS.2001-5992.164167

Dove, Michael. 2011. *The Banana Tree at the Gate: A History of Marginal Peoples and Global Markets in Borneo*. New Haven: Yale University Press.

Eilenberg, Michael. 2014. "Frontier Constellations: Agrarian Expansion and Sovereignty on the Indonesian-Malaysian Border." *The Journal of Peasant Studies* 41 (2): 157–182. https://doi.org/10.1080/03066150.2014.885433

Elden, Stuart. 2010. "Land, Terrain, Territory." *Progress in Human Geography* 34 (6): 799–817. https://doi.org/10.1177/0309132510362603

Fisher, Micah R., Daulay Muhammad Haidar, Wicaksono Satrio Adi, Bislaux Alice, and Erika Cikal Arthalina. 2023. *Forest Restoration and Rehabilitation in Indonesia: A Policy and Legal Review*. Report. EU REDD Facility.

Galudra, G., M. Van Noordwijk, S. Suyanto, I. Sardi, U. Pradhan, and D. Catacutan. 2011. "Hot Spots of Confusion: Contested Policies and Competing Carbon Claims in the Peatlands of Central Kalimantan, Indonesia." *The International Forestry Review* 13 (4): 431–441. https://doi.org/10.1505/146554811798811380

Gershon, Ilana. 2019. "Porous Social Orders." *American ethnologist* 46 (4): 404–416. https://doi.org/10.1111/amet.12829

Hapsari, Nindita, Widiastomo Triyoga, Ardila Juan, Warren Matthew, and Daniel Nepstad. 2020. *Research report: Towards sustainable and productive management of Indonesian peatlands: Case Studies of Indonesia's Peat Restoration Agency (Badan Restorasi Gambut) Interventions in Siak, Riau and Pulang Pisau, Central Kalimantan*. Earth Innovation Institution. https://earthinnovation.org/uploads/2020/08/Towards-sustainable-and-productive-management-of-Indonesian-peatlands_fin.pdf

Horton, Alex, Vili Virkki, Anu Lounela, Jukka Miettinen, Sara Alibakhshi, and Matti Kummu. 2021. "Identifying Key Drivers of Peatland Fires Across Kalimantan's Ex-Mega Rice Project Using Machine Learning." *Earth and space science* 8 (12). https://doi.org/10.1029/2021EA001873

Howell, Signe. 2014. "'No RIGHTS–No REDD': Some Implications of a Turn Towards Co-Benefits." *Forum for Development Studies* 41 (2): 253–272. https://doi.org/10.1080/13642987.2019.1579990

Huff, Amber. 2021. "Frictitious Commodities: Virtuality, Virtue and Value in the Carbon Economy of Repair." *Environment and planning E: Nature and Space* 6 (4): 1–26. https://doi.org/10.1177/25148486211015056

Jewitt, S.L., D. Nasir, S.E. Page, J.O. Rieley, and K. Khanal. 2014. "Indonesia's Contested Domains. Deforestation, Rehabilitation and Conservation-With-Development in Central Kalimantan's Tropical Peatlands." *International Forestry Review* 16 (4): 405–420. https://doi.org/10.1505/146554814813484086

Jong, Hans Nicholas. 2021. "Indonesia Renews Peat Restoration Bid to Include Mangroves, But Hurdles Abound." *Mongabay*. January 1. https://news.mongabay.com/2021/01/indonesia-renews-peatland-mangrove-restoration-agency-brgm/

Kelly, Alice, and Nancy Peluso. 2015. "Frontiers of Commodification: State Lands and Their Formalization." *Society & Natural Resources* 28 (5): 473–495. https://doi.org/10.1080/08941920.2015.1014602

Knapen, Han. 2001. *Forests of Fortune? The Environmental History of Southeast Borneo, 1600–1880*. Leiden: KITLV.

Lohmann, Larry. 2009. "Towards a Different Debate in Environment Accounting: The Cases of Carbon and Cost-Benefit." *Accounting, Organizations and Society* 34 (3–4): 499–534. https://doi.org/10.1016/j.aos.2008.03.002

Lounela, Anu and Heikki Wilenius. 2023. "Haunting the Factory: Indonesian Modernity and the Spiritual Landscape of Central Kalimantan." *Ethnos* 1–21. https://doi.org/10.1080/00141844.2023.2227773

Lounela, Anu and Tuomas Tammisto. 2021. "Introduction: Frontier Making Through Territorial Processes: Qualities and Possibilities of Life." *Suomen Antropologi* 46 (1): 5–14. https://doi.org/10.30676/jfas.v46i1.112425

Lounela, Anu. 2009. Contesting forests and power: dispute violence and negotiations in Central Java. Phd diss. Helsinki: Helsingin yliopisto.

Lounela, Anu. 2015. "Climate change disputes and justice in Central Kalimantan, Indonesia." *Asia Pacific Viewpoint* 56 (1): 62–78. https://doi.org/10.1111/apv.12084

Lounela, Anu. 2019. "Erasing memories and commodifying futures within the Central Kalimantan landscape." In *Dwelling in Political Landscapes: Contemporary*

Anthropological Perspectives, edited by Anu Lounela, Berglund Eeva and Timo Kallinen, 53–73. Helsinki: Studia Fennica Anthropologica. https://doi.org/10.21435/sfa.4

Lounela, Anu. 2021a. "Making and Unmaking Territories with Plants in the Riverine Peat Landscape of Central Kalimantan." *Suomen Antropologi* 46 (1): 15–37. https://doi.org/10.30676/jfas.v46i1.99890

Lounela, Anu. 2021b. "Shifting Valuations of Sociality and the Riverine Environment in Central Kalimantan, Indonesia." *Anthropological forum* 31 (1): 34–48. https://doi.org/10.1080/00664677.2021.1875197

Mahanty, Sango, Sarah Milne, Wolfram Dressler, and Colin Filer. 2012. "Social Life of Forest Carbon: Property and Politics in the Production of a New Commodity." *Human ecology* 40 (5): 661–664. https://doi.org/10.1215/9781478007524-012

McCarthy, John. 2013. "Tenure and Transformation in Central Kalimantan after the 'Million Hectare' Project." In *Land for the People: The State and Agrarian Conflict in Indonesia*, edited by Lucas, Anton and Carol Warren, 183–214. Athens: Ohio University Press.

MMCKalteng. 2024a. "BRG Restorasi 366.746 Ribu Hektar Lahan Gambut di Kalteng." Last Accessed May 14, 2024. brg-restorasi-366-746-ribu-hektar-lahan-gambut-di-kalteng

MMCKalteng. 2024b. "Restorasi Rambut BRG Rankul 69 Desa/Kelurahan di Kalteng." Last Accessed May 14, 2024. restorasi-gambut-brg-rangkul-69-desa-kelurahan-di-kalteng

Neale, Timothy. 2022. "Interscalar Maintenance: Configuring an Indigenous 'premium Carbon Product' in Northern Australia (and Beyond)." *The Journal of the Royal Anthropological Institute* 29 (2): 306–325. https://doi.org/10.1111/1467-9655.13861

Nygren, Anja and Anu Lounela. 2023. "Remaking of wetlands and coping with vulnerabilities in Mexico and Indonesia." *Water Alternatives* 16 (1): 295–320. https://www.water-alternatives.org/index.php/alldoc/articles/vol16/v16issue1/690-a16-1-7/file

Peluso, Nancy. 2012. "What's Nature Got To Do With It? A Situated Historical Perspective on Socio-Natural Commodities." *Development and change* 43 (1): 79–104. https://doi.org/10.1111/j.1467-7660.2012.01755.x

Peluso, Nancy. 2017. "Plantations and Mines: Resource Frontiers and the Politics of the Smallholder Slot." *The Journal of peasant studies* 44 (4): 834–69. https://doi.org/10.1080/03066150.2017.1339692

Polanyi, Karl. 2001 [1944]. *The Great Transformation*. Boston: Beacon Press.

PP - Peraturan Presiden Republik Indonesia No 2 Tahun. 2015. *Rencana Pembangunan Jangka Menengah Nasional (RPJMN) 2015–2019*.

Rasmussen, Mattias Borg, and Christian Lund. 2018. "Reconfiguring Frontier Spaces: The Territorialization of Resource Control." *World Development* 101: 388–399. https://doi.org/10.1016/j.worlddev.2017.01.018

Regjeringen. 2024. "Indonesia and Norway strengthen joint efforts to beat climate change." Last Accessed May 22, 2024. https://www.regjeringen.no/en/aktuelt/indonesia-and-norway-strengthen-joint-efforts-to-beat-climate-change2/id3016606/

Resosudarmo, Ida Aju Pradnja, Stibniati Atmadja, Andini Desita Ekaputri, Dian Y. Intarini, Yayan Indriatmoko, and Pangestuti Astri. 2014. Does Tenure Security Lead to REDD+ Project Effectiveness? Reflections from Five Emerging Sites in Indonesia. *World Development* 55: 68–83. https://doi.org/10.1016/j.worlddev.2013.01.015

Ribot, Jesse. 1998. Theorizing Access: Forest Profits Along Senegal's Charcoal Commodity Chain. *Development and Change*. 29: 307–341.

Sari, Agus P., Alue Dohong, and Budi Wardhana. 2021. "Innovative Financing for Peatland Restoration in Indonesia." In *Climate Change Research, Policy and Actions in Indonesia*, edited by R. Djalante, J. Jupesta and E. Aldrian. Springer climate, Springer. https://doi.org/10.1007/978-3-030-55536-8_12

Schiller, Anne. 1997. *Small Sacrifices: Religious Change and Cultural Identity among the Ngaju of Indonesia*. New York: Oxford University Press.

Sheridan, Michael. 2016. "Boundary Plants, the Social Production of Space, and Vegetative Agency in Agrarian Societies." *Environment and Society* 7 (1): 29–49. https://doi.org/10.3167/ares.2016.070103

Shimamura, Tetsuya. 2017. "An overview of Tropical Peat swamps." In *Catastrophe & Regeneration in Indonesia's Peatlands: Ecology, Economy & Society*, edited by Kosuke Mizuno, Motoko S. Fujita and Shuichi Kawai, 123–147. Singapore: NUS PRESS in association with Kyoto University Press.

Tsing, Anna Lowenhaupt. 2005. *Friction: An Ethnography of Global Connection*. Princeton: Princeton University Press. https://doi.org/10.1515/9781400830596

Tsing, Anna Lowenhaupt. 2015. *The Mushroom at the End of the World: on the Possibility of Life in Capitalist Ruins*. Princeton: Princeton University Press.

UNDP. 2024. "Support facility for the peat restoration (badan restorasi gambut)." Last Accessed May 28, 2024. https://www.undp.org/indonesia/projects/support-facility-peat-restoration-badan-restorasi-gambut

United Nations Environment Programme. 2009. *Blue carbon: the role of healthy oceans in binding carbon*. https://wedocs.unep.org/20.500.11822/7772

UNOPS. 2024. "Restoring indonesian peatlands, protecting our planet." Last Accessed May 14, 2024. https://www.unops.org/news-and-stories/stories/restoring-indonesian-peatlands-protecting-our-planet

USAID 2020a. Final report USAID LESTARI 2020. https://pdf.usaid.gov/pdf_docs/PA00X1BK.pdf

USAID 2020b. USAID LESTARI: Lessons learned technical brief. Free, Prior and Informed Consent: A Tool for Improved Land Use Governance. https://www.climatelinks.org/sites/default/files/asset/document/2021-04/2021_USAID_LESTARI_FPIC-A-Tool-for-Improved-Land-Use-Governance.pdf

Vandergeest, Peter, and Nancy Lee Peluso. 1995. "Territorialization and State Power in Thailand." *Theory and society* 24 (3): 385–426. https://doi.org/10.1007/BF00993352

Vergouwen, J. C. 1921. "Tatah- en Soengeirechten." *Koloniaal Tijdschrift* 10: 545–562. https://doi.org/10.1080/00664677.2021.1875197

7 La#oma's pre- and post-carbon landscape*

The ont*-politics of a vanished village

Ingmar Lippert

Introduction

How were people in the Global North able to ignore climate change for so long? Wasn't carbon politics ubiquitous? Such questions might be asked by a future archaeologist. I propose that one reason is that in the early 2020s, our imagination of carbon is still limited by the stories we tell and read about carbon and how these stories are explicitly or implicitly saturated by knowledge and ignorance. Even more tricky, these stories are not merely issues of words but of materiality, too: stories are also materially infrastructured, con-figured. In this chapter, I seek to investigate the infrastructures that con-figure stories of the experience and remembering of carbon in a village destroyed for lignite mining and speculate about capacities for alternative stories of carbon and that village. In that line of investigation, we encounter, *inter alia*, information panels and a digital display that provide competing accounts of the relation between carbon and the village.

While sustainability crises are present (Mårald and Priebe 2021) and unsustainabilities are sustained (Blühdorn 2007), a range of competing transformations are underway. Which transformations to withstand climate crises can we imagine? The social studies of science and carbon economics show that carbon politics involve issues ranging from climate change, emission trading, the outsourcing of pollution and denial to the politics of knowledge (Sluijs et al. 1998; Lohmann 2005; Ninan 2011). Even though carbon is one kind of entity, it exists differently in different places, and it is practically brought into being within situated realities, effecting carbon-as-multiple (Lippert 2015). We come to know about carbon as it is evoked explicitly across various social, cultural, political and economic world-making endeavours (O'Reilly et al. 2020). Accounts and stories of carbon emerge in a universe of references to carbon (Lippert 2011a), not only in explicit ways. *Implicitly*, carbon "con-figures" (Suchman 2012) stories of the everyday, of the past and present: carbon is one of many entities that are woven into specific threads that shape stories. And such stories are located; they are not abstract but are performed in-place. This

---

* The hash serves as a placeholder of multiple spellings of the village, and stories associated with the spellings. This intricately formed spelling of the village is discussed alongside unfolding the chapter's argument. See also note 5. For the star in ont*-politics, see note 9.

DOI: 10.4324/9781003478669-8

chapter investigates distinct modes of storying the transformation of a landscape and its village into a place of ongoing, multiple and contested existence. The village is Lakoma (a German spelling, though within the village I also find other spellings; see next section). It is part of the municipality of Cottbus, around 125 km upstream of Berlin, Germany. I argue that storying constructs the place as one of carbon mining, a place against carbon mining, a place of post-carbon mining and a place for and against the green transformation. I ask what the substantive grounds might be for a mode of storying within a landscape of transformations that gives voice to contestation and non-coherence, conflict and non-commensurability, between natures, cultures and carbon.

In this chapter, I mobilize "carbon" as a prism. As a prism, I, and (other) environmental activists, bring it to the place in question and invest in problematizing how human relations to carbon shaped the place. I attend to a case that is entangled with accounts of lignite mining. Carbon stories involving lignite mining are complex. While lignite, as a form of brown coal, is a particularly dirty fossil fuel (Braunbeck 2020), mining involves jobs and profits too. In our case, we encounter opencast mining, a form of extractivism that leaves ruined landscapes. The local and immediate effects of this mining can be seen by the human eye and do not require the type of statistical mediation and computational constructions that characterize hegemonic concepts of carbon (Lohmann 2005; Lippert 2015). Facing these ruins, I show how within a mining landscape and a village located there, extractivist commitments are sustained, legitimized and hidden.

Zooming in on Lakoma

A warm summer evening in 2022. A circle of colourful chairs, benches and other things to sit on provides a temporary space for a heterogenous crowd of villagers, if I may call them so. They lived, or dwelled, in Lacoma, at some point between 1990 and 2006. I was part of this group because Lacoma, just outside of the city of Cottbus (Germany), was a place that I became very fond of when I studied an undergraduate degree course in environmental and resource management there in the early 2000s.[1] In Lacoma, I encountered villagers who shaped the village's rituals, routines and daily life collectively; I met fellow students, who were invested in low impact life experiments, seeking to live sustainably, working with wood, strategizing environmental politics in the small and in the big. Environmental politics was one of the village's defining topics, but so too was alternative life, for a range of punks and counter-cultural folks who had made Lacoma their home. In the 2022 circle, I heard about a multitude of ways of living in and bringing life into Lacoma. That these mattered was significant because between 1987 and 1990, most villagers were quite forcefully pushed out of the village (Weber and Cabras 2017) to create space for lignite mining. With the collapse of the German Democratic Republic (GDR), energy consumption patterns rapidly shifted (Schleich et al. 2001). The opencast mine next to Lacoma did not expand as quickly as expected, not swallowing the village in 1990, but 16 years later. In this interstitial temporal space, Lacoma was occupied by different people and groups, some of whom joined the 2022 summer circle and the Lacoma festival that opened the next day. In that circle, I witnessed myriads of accounts of living in Lacoma. They were deeply personal – full of wonder, laughter and tears. The circle evoked not one, but multiple Lacomas.

Recognizing multiple Lacomas, this chapter is committed to studying sideways (rather than down; see methodology section below). I explore the presentation and inscription of Lacoma between 2018 and 2022 and the affordances of experiencing Lacoma in that phase. This means I focus on devices and infrastructures of inscription – such as information panels, digital displays and monuments – that either represent Lacoma and/or take part in constituting realities of Lacoma (including but not limited to heritage infrastructure), while narrowing the scope to sites which can be explored on foot: I explore Lacoma within and around the village as well as within a museum that tells of Lacoma.[2]

Given that Lacoma is a village that anticipated destruction as a result of accessing carbon – a village that experienced destruction, causing carbon emissions – I wonder how contemporary inscribed realities of Lacoma – in monuments, museums and other places – relate to carbon, especially with respect to regional concerns within the so-called "Transformation" aka "Strukturwandel" (structural change) or "Kohleausstieg" (coal phase out), a transition to a post-mining economy in the region.[3] What transitions do Lacoma's relations to carbon indicate?

My concern with what or how Lacoma is, and how it is socially-culturally brought about and transitions, resonates with Norgaard's (2018) call to imagine carbon sociologically. With Norgaard, I mobilize Mills' (1959) interest in how personal troubles relate to seemingly larger societal issues. In *The Sociological Imagination*, Mills suggests we have the "capacity to range from the most impersonal and remote transformations to the most intimate features of the human self—and to see the relations between the two" (1959: 7). Mills indicates that the researcher/subject may set out from a vantage point of their biographical position "to grasp what is going on in the world, and to understand what is happening in themselves as minute points of the intersections of biography and history within society" (ibid). To imagine Lacoma sociologically, I operationalize this form of relating by presenting my encounter with Lacoma from the vantage point of a situated and overtly politically interested subject (now concerned about climate politics, in the early 2000s taking part in direct action protests in Lacoma, concerned with Lacoma, not carbon). I read inscriptions in and of Lacoma. I analyse these inscribed realities of the village, and their situated readings/read-abilities, as shaping how regional transformation as well as Lacoma are, and can be, both known and contested – at this intersection of biography and history within society. While I share Mill's classical sociological motivation, analytically, in this chapter, I am informed by material-semiotic forms of sociology that have shaped the field of Science and Technology Studies (STS) (see Law 2008). It is from STS that I borrow a nuanced repertoire of analytical terms. This includes Suchman's (2012) configuration, but also inscription (see for instance Latour 1993), apparatus (referring to both my research design and practice, as well as to the knowledge-making machinery that creates knowledge about La#oma; see Barad 2007); and in "STSy" forms of slightly playful writing, I engage both with human subjects (knowing about the village) and with non-human entities (like the village) that become subject to human practices.

In line with this inspiration from STS, I train my work of imagining Lacoma specifically with feminist technoscience studies insights into the emergence of realities, shaped both materially and semiotically. This is the core of my analytical approach. While it is quite intuitive that knowing a world and acting in or on it may produce reality effects, in STS we need to foreground: the world is made practically, bodily or otherwise materially or infrastructurally, and that may or may not involve knowledge. In the literature on so-called ontology, I find analytical resources that sensitize me to the material entities involved in configuring users, actors and other humans in knowing the world (Suchman 2012). Part of such processes of configuring knowledge may be both (a) ontologies, more or less formalized, and (b) non-theorized commitments to the world being there, being in this way and not in others (Verran 2001; Mol 2002; Barad 2007). To contour these analytical sensibilities, I continue experimenting with the neologism ont*-politics, which promises a nuanced analysis of reality effects not necessarily determined by formal ontologies (Lippert 2020).

One way in which I will argue that ont*-politics is visible empirically is through the competing accounts of the destruction of the village as achieving a seemingly non-commensurable and highly different politics of transition. That Lacoma's reality is thoroughly contested also becomes apparent when we turn to the designating or spelling of the village's name. Lacoma, in fact, is the primary spelling currently mobilized by a distinct community – activists who were invested in saving the village and its environment from mining, those seeking to prevent the brown transformation. That spelling is illustrated by Michel (2005), but it is in fact older. The village authorities in the GDR referred to the village as Lacoma up until the 1970s (before that, it was also known as Lacuma). Another spelling, Łakoma, is Sorbian, a Slavic language spoken by a minority people in Germany who claim autochthonous status for the region called Lusatia, to which Łakoma belongs.[4] The village was predominately Sorbian until 1956 (AvO 2010, 100). Subsequently, the village was increasingly Germanized. In German, the village is called Lakoma, and as such, it is spelt this way in most literature on the village. In what follows, I refer to the village as *La#oma* to imagine its multiple potentialities.[5] Analytically, this potentiality of the village's reality allows us to identify its varying ont*-politics: the non-deterministic ways of imagining, knowing and enacting La#oma.

Across the broad literature on energy, environment and society relations, La#oma is a site that is evoked, *inter alia*, for several purposes. To some, it is an indicator of how long the GDR continued to displace people for lignite mining (Weber and Cabras 2017); to others, it represents a case of the mobilization of Sorbian cultural heritage against industrial dispossession (Hagemann 2022) or an exceptional case of the GDR security service (Stasi) reporting on the villagers (Soch 2020, 99–100), as some of these were seen as radical because they staged a protest in Berlin against their dispossession (Reaves 1996). The village has also been considered relevant for local cultural heritage, which needed to be documented before its destruction (Förster 1998; AvO 2010). Beyond public and corporate research into local geology and mining opportunities, La#oma was also subject to natural science monitoring, given that several highly protected species lived there (BfN 2024) and the design of compensation measures that sought renaturation after its destruction (Pflug 1998; Müller et al. 2015). La#oma is shaped by carbon and by contested values of carbon; it has been discursively and materially mobilized to achieve a brown transformation and to imagine a green transformation.

Today, La#oma exists in several places simultaneously, not just in studies and other stories, but also in forms that allow visitors to walk within or across the village. I turn to two such walkable sites, the site of the now mostly destroyed village at the geographic coordinates of La#oma and the spatial place called Lakoma (approximately lat. 51.7964312, lon. 14.3815733), as well as Łakoma in a museum, the *Archiv verschwundener Orte* (which I translate as Archive of Vanished Villages, but below I note an alternative translation), located about 24 km away from these coordinates in another town, Horno. And, be warned, within the former sites, La#oma exists multiply, too.

With this minimalistic multisited approach, I attend to how La#oma is and how, in its multiple forms, competing transformations are foregrounded. My narrations of La#oma draw on a corpus of empirical materials that allows me to attend to being and difference within La#oma's presence. Politically motivated, I wonder how these versions of the village enable, disable or otherwise prefigure conversations about regional transitions and energy transition. In following this wondering, I show myself as a subject of the village in conflicted configurations, and I show the village as not merely a place of the past but as a site of ongoing politics. I problematize how some organized epistemic and ontological forms of naturescultures seek to dominate others. My wondering leads me to trouble contemporary heritage infrastructure that inscribes stories of La#oma: I speculate about heritage infrastructure reconfigurations that might foreground the contested naturescultures of La#oma and their role in regional transformation discourses. On these grounds, this chapter argues that across the La#omas – that is, in contemporary modes of infrastructuring the remembering of the village and its environs – a rich set of histories and realities are mobilized in troubling and significantly limited ways. Taking into account this rich set might allow us to think of better ways to configure the (heritage) infrastructures for remembering the multiple La#omas and their contested environmental, energy and carbon relations, as well as their complex promises for transitions.

To develop the overarching argument that the multiple histories and realities of the village are not (yet) well used to power an emancipatory energy transition, let alone a transition to sustainability, I first present my analytical apparatus, then sketch my methodological approach and subsequently,

I narrate a story of La#oma. To achieve the latter, I provide snippets of my encounters with La#oma, by ethnographically reading and by walking/experiencing the village at the two distinct sites. I analyse my encounters in terms of the ont*-politics these indicate and conclude by considering the implications of this for heritage politics in a region characterized by competing carbon stories.

Analysing naturescultures as ont*-politics

With the neologism of ont*-politics, I open my sociological imagination to an analysis of practices that enact the multiplicity of naturescultures in non-deterministic ways. In this analysis, ontologies may matter – but whether, how and to whom they matter is not guaranteed. Before delving into this conceptual knot, let me guide you there via my interest in the conversation on naturescultures.

The presentation of La#omas in the empirical part of this chapter indicates both references and realities that seem neatly divided into natural conditions and entities versus cultural ones. However, my empirical material will also indicate realities that do not sit easily exclusively on either side. For instance, the ever-present lignite, a matter of organic carbon and geological nature, is partially mobilized discursively and culturally with a naturalizing claim of mining as an antecedent necessity. To develop a nuanced analysis of things and relations between natures and cultures, here I draw on STS and anthropology. I then return to the empirical in subsequent sections.

Anthropology has developed highly generative analytics of the relational constitution of natures and cultures in conversation with STS (Tsing et al. 2017; Gesing et al. 2019). Latour (1993) recognizes the relativist and plural understandings according to which different cultures can know the world differently. However, he also recognizes that the anthropologist might analyse Western natures, beyond classical concerns in anthropology, with "other" peoples' understandings of their natures and environments. Of course, Latour's imagined anthropologist is trained in a modern thinking that separates nature from culture. From actor-network theorist Latour, I draw the concept naturescultures to highlight that not only cultures can be achieved in diverse ways, but natures, too, *both potentially and in the plural*. "Naturescultures" suggests that there is no such thing as a self-evident cut between "nature" and "culture"; instead, any such separation has to be achieved by human labour, for instance, in laboratories. I draw specifically on the capacity of this conversation to go beyond the classical concern in STS with laboratories and engage here with a village, its landscape and its natures. In this engagement, I maintain an interest in how complex village-landscape-natures are known (the epistemological concern) and what that complex is and comes to be (the ontological concern). I return to ontology and ontics in more detail in a moment, but first I attend to the significance of place in my analysis. This generates my sensibility to situated and located achievements of naturescultures that may be highly contingent:

such achievements generate all kinds of naturescultures, including forms of knowing and enacting some in which nature and culture are separated out from each other.

That place matters in the constitution of knowledges and realities is not only widely recognized in anthropology and human geography, but this is also demonstrated in STS engagement with the constitution of realities. Mol (2002) shows that at different places the same object can be enacted differently.[6] Barad (2007) similarly provides an analysis that underwrites the relevance of place and the role of gender and habitus that are prevalent in specific sites.[7] We can expect specific entities to shape a place. Reading technoscientific entities as "figures", Suchman (2012) asks how these entities are "con-figured": how are a range of material, embodied and imaginary objects related to achieving certain realities? In this chapter, I separate the "con" (aka "together" in Latin) with a hyphen to underline that in con-figurations, several figured entities are related. I see con-figuring as a relational process in which figures and entities that make up con-figurations shape each other non-deterministically. With this sensibility, STS can help us understand not only technoscientific objects as they appear in physics research (Barad 2017) but also digital enactments of environments (Lippert 2015), and it can build on older accounts of the emplaced con-figuration of places, including in colonial settings (Verran 1998; Sioh 1998).

STS accounts for these enactments of realities across registers of knowledge and being, the episteme and the ontic, epistemology and ontology. In the actor-network-theory reading of classical social constructivism, the world can be represented and known differently, and matter – material reality (found in technology, human bodies, trees, etc.) – is left untouched. In a distinction between knowing an independent reality "out there", Mol (2002) points out that some knowledge practices also alter reality, for instance, by cutting a body. For her, which realities are enacted, where and how literally matters: materially, a body with atherosclerosis is enacted as existing as "more than one—but less than many" (55). It matters, too, politically. Different goods can be enacted in practices. For Mol (1999), then, enacting specific realities is political, a practical question of ontological politics. In her research on physics, Barad (2007, 163, 2017) foregrounds that knowing and materially con-figuring realities appear entangled. Underlining this entanglement, Barad develops the concept of onto-epistem-ology. In the scientific apparatus, phenomena are onto-epistem-ologically con-figured. A slightly different analytical take on the study of practice in which knowledge and being are constituted appears in Verran's work. Ontic, in her work, refers to "the level of entities' existence or being" (Verran 2007a, 34) – that is, realness, where entities are to be understood as actor-networks that are achieved and performed in material-semiotic practices that include our (not necessarily explicit) commitments to these entities being there. In Verran's analytical distinction, ontology – while also being performed and, thus, shaping ontics – is characterized by explicitly explaining, studying or theorizing what is and/or the metaphysical commitments to the being. Ontology and ontics, in short, do not necessarily determine each other.[8] Why engage with this nuanced conceptual discussion? I draw on Verran's work because it provides me with the critical capacity to analyse heritage infrastructures of or in La#oma that may achieve material as well as a semiotic shaping of carbon discourse, while the entities mobilized (in said discourse) are not strictly and formally determined by ontologies.

To mobilize this analytical space, I recognize that the latter three scholars' various ont* concepts both differ and resonate.[9] I attend to practices and infrastructures involved in naturescultures and enquire into how naturescultures are known and how the processes of knowing shape each other (natures, cultures and knowledges), without presuming that knowing determines what naturescultures become to be. This allows for a (non-linear) continuum between action with reality effects that get ontologized, and action that has reality effects without being guided by a specific ontology (Raasch and Lippert 2020). An analysis of the ont*-politics of a phenomenon is useful to analyse the manifold more or less ontologically informed practices that achieve the reality effects that shape that phenomenon. The practices and infrastructures may well effect realities that are firm and stable, touchable (as in Mol's (2002) work on human bodies); however, other realities may be much less firm, quivering in and out of the real as we typically know it (consider Barad's (2017) take on virtual particles in physics), effecting ghostly phenomena.[10] Ont*-politics is a "politics of what" (Mol 2002) that is not necessarily theorized or represented by specific knowledge forms but that is shaped by the inseparability of what is and how that being is imagined, known and/or committed to, while neither determines becoming.

How we tell stories of the making of natural cultural realities requires care because our stories relate to discourses, can shape these, intervene in them or strengthen their stability, support us or get into our ways (Swanson et al. 2017). Of particular importance here are those voices that get forgotten and whose realities are erased (Gan et al. 2017, G6; Morini 2019; Braunbeck 2020). But beyond that, with Verran (2007b, 2012), I suggest our focus needs to be attuned to how stories of infrastructures can take part in ont*-politics, rendering, for instance, the destruction of some, or the stabilization of other, environments and knowledge forms.

Materials, methodology and multi-modal storying

In this chapter, I attend to two sites with a form of attention that generates a particular analysis, an analysis that is partial, situated and located.[11] I focus on sites from which I can experience La#oma through multiple modes: listening, talking, reading, walking and more. I analyse inscriptions and inscription devices and infrastructures for their con- or prefigurative role in stabilizing stories about, as well as realities of, La#oma. This research strategy respects the regional as well as activist scepticism towards interviews and qualitative social science research methods that are linked to experiences with police and secret services in the GDR and beyond (which I have been told about repeatedly by people who lived in Lusatia). While I did talk to people in this research, I did so rarely and primarily investigated the infrastructures of stories.

The first site is a museum that calls itself an archive – Archiv verschwundener Orte – Archiw zgubjonych jsow. Cunningham (2013) translates this as the Archive of Disappeared Places, but I translate it as the Archive of Vanished Villages, for this translation highlights villages, not places (which may enfold so much more than villages). Here, I simply call it the Archive. It is located in Neu-Horno, part of the city of Forst, on the eastern border of Germany, neighbouring Poland.[12] The Archive's construction in Neu-Horno is the result of a political compromise between four actors: the energy corporation Vattenfall, which owned the mining sites at the time Horno was devastated; the people of Horno, organized as the foundation Stiftung Horno; Domowina, as Vattenfall's partner and as the de-facto regional representative of Sorbian interests, given that some of Horno's inhabitants identified as Sorbs; and the municipality of Forst. Formally, the overarching objective of the Archive is the collection and scientific analysis of materials on the devastation of villages and their resettlement in the Lusatian "lignite mining district" (AvO 2010, 26). In the publication AvO (2010), we find two prefaces written by representatives of Horno and Domowina, respectively, in which we learn that the Archive is a site of reminder and reflection, and it provides information on how the solution of "socially acceptable" resettlement was realized (Vattenfall preface). Virtually, the Archive is powered by a database. The database stores "information and photos" of "all the 137 villages that had to completely or partially give way to lignite mining" (AvO, n.d.). These villages are referenced on an enormous map on the ground, visible in Figure 7.1. One of the villages databased is La#oma. I visited the Archive five times between 2019 and 2022, mostly for several hours, conducting reflexive video ethnography and taking notes on interactions I observed or took part in and conversations I overheard. On several occasions, a research assistant, and another time a colleague, accompanied and helped me to record experiences at the Archive (Figure 7.1).

The second site is Lakoma, next to the city of Cottbus, about 24 km away from the Archive. To analyse my experience of this place, while being inspired by decades-old recollections, here, I primarily draw on two ethnographic days of walking in and observing the place and its people, as well as my experience of the place. On one occasion, a colleague accompanied me. Figure 7.2 is a map of Lakoma from 1921 that shows the village's position between two types of water bodies (marked in blue), the Hammergraben canal (later called Old Hammergraben) and various ponds in the northeast. Figure 7.3 is a contemporary map, showing "Lakoma - Łakoma" next to a large human-made body of water, the Cottbuser Ostsee (I briefly return to its history below). Lakoma is now located between the New Hammergraben Canal (thus, New Hammergraben, which was constructed a few decades ago; Old Hammergraben was constructed in the 15th century) and that lake.

La#oma's pre- and post-carbon landscape 163

Figure 7.1 Archive: view of its single exhibiting room.
Source: Photograph by author.

Figure 7.2 Map: Historical view of Lakoma.
Source: © SLUB Dresden / DDZ.

164 *The Cultural Complexity of Carbon*

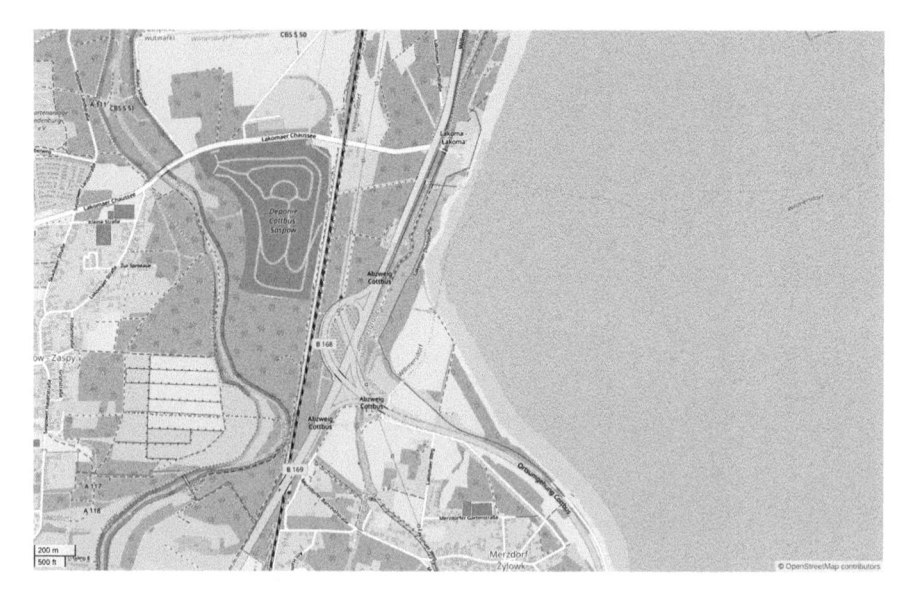

Figure 7.3 Map: Contemporary view of Lakoma in its regional context.

Source: OpenStreetMap data is available under the Open Database License version 1.0.

Experiencing La#oma

In the Archive

As an introduction to the Archive, staff point me to a welcoming video. The five-minute presentation, dated 2017, renders several things recognizable: the mining activity in the region provides over 10 billion tonnes of lignite to be burned; framed as a "mining region", the video presents lignite as an economic resource.

Leaving a vestibule-like space, I enter the Archive's main space by taking a step onto its carpet, which is a map. With curved and continuous transitions into the walls, the map connects the Archive's centre, its margins and the stations and it centres on the German part of Lusatia (see Figure 7.1). On the map I find borders, railways, streets, rivers, major lakes and reservoirs of lignite; the map also marks villages and cities. Disappeared villages are marked in red. The map's legend dates the carpet – 2006 – and details seven kinds of opencast mining as well as the locations of special waste and (partially) devastated villages.

I search for La#oma in the Archive; I find it on the map, but it only appears as a red mark. Besides viewing and literally walking on top of La#oma when I walk on the map (Figure 7.4), I cannot find out much more about the village. On one wall of the room, I find artefacts and references to La#oma. A windowed display case presents a mug with two inscriptions, "1594 Lacoma" and, below, "1989 Lacoma". A note states that this mug was given to the villagers

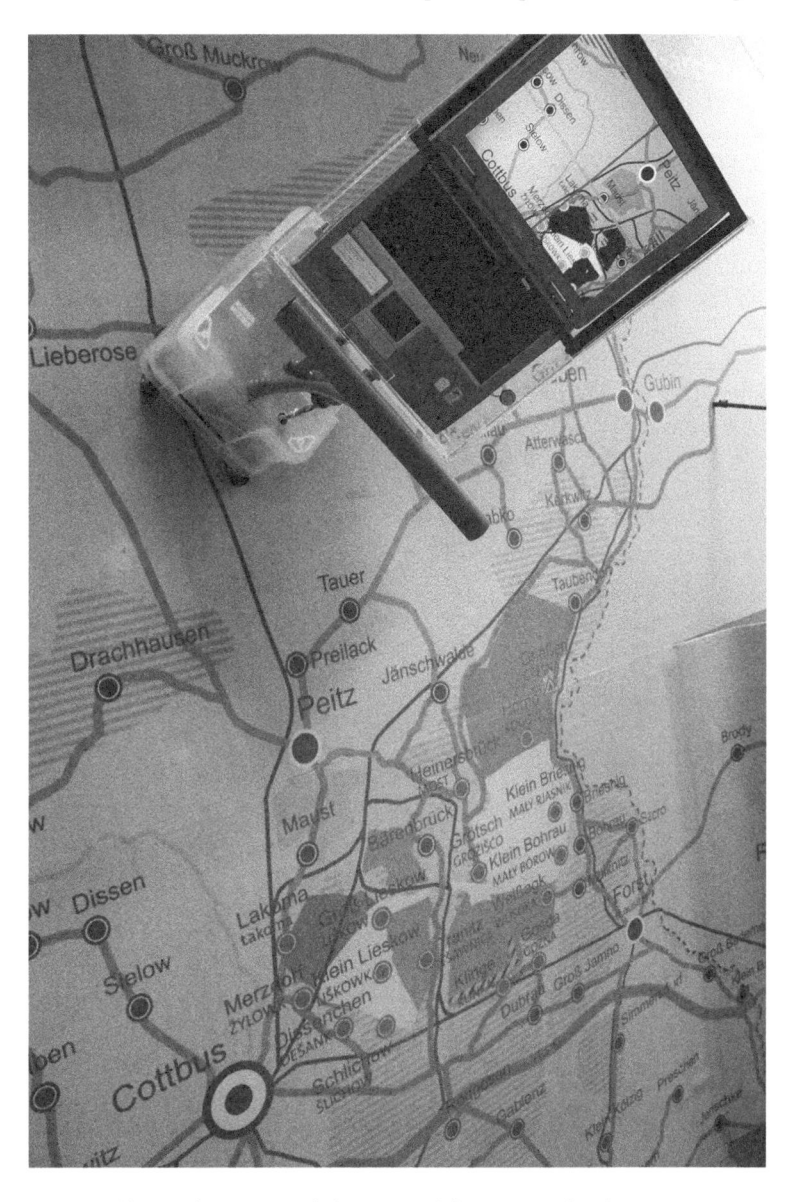

Figure 7.4 Archive: Infosauger and the mapped floor around Lakoma.
Source: Photograph by the author.

of Lakoma to commemorate their resettlement in 1989. The note explains that Lakoma "had to" be devastated to make way for the Cottbus-Nord mining site. It also tells of an "insurrection" of five women from Lakoma who travelled to Berlin in 1984 to protest against the "devastation", which was announced in 1983. Furthermore, the note mentions a Dr. Kühn (born in 1936), who recalls the initial public announcement. To the left of the case, a button allows us to listen to a five-minute audio recording of Dr. Kühn.

I turn to a device that one of the staff has activated for me, a so-called Infosauger (Figure 7.4). It has been waiting to be used in a corner of the Archive and offers a touchscreen. I can steer the device across the room by pushing its red handlebar. On the touchscreen display, I see a part of the map of the location above which the Infosauger hovers. The display captures my attention, and everything else around me ceases to matter. The displayed map follows my movement with the Infosauger, as I push it across the carpet, towards Cottbus, while the displayed map updates stumblingly to represent the map across which the Infosauger hovers. I stop once I see what I have been looking for. I tap on the display, on the red mark that indicates Lakoma/Łakoma village.

With this tap, the Archive's database opens – "Loading" the touchscreen tells me, and then shows four stacked record cards in warm earthly colours: ochre, red, brown and the text in a black font. The cards are labelled resettlement, village pictures, culture and language and local facts. The top card, "local facts", shows a photo labelled "Lakoma/Łakoma, aerial image 1987". Below the image, a menu allows me to access more details within the local facts category. I tap an image of a table that reveals a tabular presentation of eight "facts", including the name of the devastated village (Lakoma), the causal agent of the devastation (opencast mining Cottbus-Nord), the date of devastation (1987–1989 partial devastation, 2003–2006 daily press), the degree of devastation (complete), the number of resettled people (143) and three more legal-administrative categories (e.g. the "region" to which the village belongs, here "Lower Lusatia"). The menu provides a compass, which I select and then read a sentence on the location of Lakoma relative to Cottbus. Then the menu provides a pile of books, which I select, too, and which reveals some literature sources on that village, including a reference to a co-authored report by the environmentalist organization Grüne Liga and a local critical citizen initiative that works with the ponds around Lakoma.

Except for the village pictures, the other four record cards provide primarily various factual social accounts, contextualized in terms such as the "cultural", "political", "administrative" and extending to the "religious". I learn about Sorbian costumes, language and festivities; the volunteer fire brigade; and a school. I read about major historical events (e.g. a war that destroyed the Lakomaer Teiche ponds in 1756–1763 and a flood that destroyed most farmlands in 1771) and phases (village dwellers conducted fishery work in the ponds, and in 1777, following Prussian king Frederick the Great's orders, potatoes were cultivated). The pictures category presents mostly combinations of sociocultural artefacts (such as streets, houses, power poles, lanterns and gardens).

Within the "culture and language" category, I found a card on festivities, which provides information about customs like erecting a maypole, adorning yard gates with green birch twigs and capturing roosters. A card on "structures and places" names 21 water bodies, including the ponds. The names jog my memory of the early 2000s: I am reminded of going for walks with Lacoma-based environmental activists between the ponds or enjoying a bath in the Old Hammergraben canal. The card on associations shows a horse-drawn fire brigade carriage, leading a procession. The cards displaying "culture" reference entities and relations that I will analyse as naturescultures. The "village pictures" category lets me glimpse environments and landscapes, for example, a forest or grass verges with a stork's nest in the background. "Recultivation" is the final topic of the "resettlement" category. That category ends with a photo of an information board titled "The lignite opencast mine Cottbus-Nord on way to the Cottbuser Ostsee" that visualizes an expected future 1900 ha (19 km^2) large artificial lake, which is a key part of the water-infrastructural imaginary of a "Lusatian Lake District".

At the geographical site of La#oma

Setting out to visit La#oma at its geographical, "original", site, I look for a public transport connection and find in my smartphone app a bus stop named "Lakoma". From Cottbus University campus, I walk to a bus station and board a bus that supposedly goes to La#oma. I ask the bus driver for a ticket to that destination. They do not understand me. I have to repeat the name, and then I get the ticket. Once I sit down, it gets exciting. A display shows the upcoming stops. After some minutes, Lakoma appears on the display, and soon we arrive at a bus stop that I vaguely recall from my time in the village 20 years ago. I get off the bus. Outside is a pole with a sign reading "Lakoma". Several bus timetables are fixed to it, as well as a bicycle.

The pavement points me in one direction; I walk on and see a number of white wooden crosses on the ground (Figure 7.5) to my left. I count 13 crosses, each marked with the name of a village in German and Sorbian. Death is looming. Yet, among the crosses is an artistic windmill installation: a rusted rod, about three meters high, with four asymmetrically positioned wings of different colours and patterns: blue, green, orange and white. I wonder if this is a dream for an alternative to death.

I continue my walk, noting road signs with directions to Cottbus and its various districts, including Lakoma. Each is also named in Sorbian in small script. A signpost for cyclists does not mention La#oma, but it does mention the Cottbus-Nord mine. I turn left towards an approximately 15-m-long bridge that crosses the New Hammergraben canal. Just before the bridge, on the left, I see a double information panel. I take photos and move on. On the left-hand side, I see a homestead, a house with a neighbouring barn. The next seconds, as I move towards the bridge, cross over and then come off it, shatter me. The movement feels utterly familiar, an embodied and emplaced routine that triggers an expectation of what view I should see from the bridge, rooted in many years of visiting La#oma in the past – a village, ponds and trees in the background. But today that expectation is not met. I see a parking lot behind an embankment, which cuts off a wider view of the horizon.

168 *The Cultural Complexity of Carbon*

Figure 7.5 Left before the bridge across the New Hammergraben to La#oma: 13 white crosses and the artistic windmill.

Source: Photograph by the author.

Before the embankment, I see another information panel, next to a large tree, seemingly the last large tree standing (others had been felled), many of its leaves turned yellow. Next to the tree is a small sign, declaring this site as a fire brigade functional site No. 7. On the embankment I see bushes, grass, a bench with a wooden roof and a footpath leading there. Next to that footpath, I see a board advertising snacks, available on weekends.

I arrive at the information panel and find it to be three-sided, offering an (over)load of information. I cannot identify an author, but a source is provided: a document about Lakoma village published by the mining operator in 1988. One side of the panel with the heading "Lakoma and mining" provides a brief history, covering two periods. I learned from a 1594 to 1615 report about the structure of the early settlement and then about the period from 1976 to 2030, in which Lakoma and its environs were first declared a prospective mining site and then the announcement of the largest Lusatian lake "Cottbuser Ostsee". Within the latter period, in 1983, the need to devastate Lakoma was declared, followed by a brief phase of protest in which villagers directly addressed the GDR ruling party's (SED) central committee. By the end of the 1980s, the villagers had sold their land. The remaining period is shaped by compensation planning, marked by a 2007 decision for the actual devastation and for archaeological research to begin within the village. A second side notes three spellings of La#oma, that is, Lacoma, Lacuma and Lakoma, recognizing that it was not until the 1970s that the villages adopted the Lacoma spelling; Lacuma and Lakoma were both used already since the 16th century. The main text recognizes that Łakoma was the denotation used in the first documented administrative recognition of the settlement. According to the text, the village is a neighbour of the (Old) Hammergraben canal, which provided water masses to power the nearby military industry in Peitz, but the man-made stream also irrigated the pond area. A historic map shows Lacoma, marked in red, the Hammergraben and Peitz. An overlay to the map depicts a mug, an image of the one I had seen in the Archive. Interestingly, the third side, titled "Lakoma and the ponds", provides alternative theories as to the origins of these ponds. One suggests they originate in 16th century (or earlier) fishery practices, another in the 18th century. The panel's margins indicate "close to 24 ponds" that once existed. The panel suggests a "prudent" management of the ponds that resulted in a "valuable natural space". Subsequently, the panel introduces the compensation measures with their objective to achieve renaturalization by way of relocating the River Spree 11 kilometres into its "original" riverbed. From a tourist perspective, the measures are considered successful. The panel then details the number of newly created compensation ponds (8), their spatial dimensions (21 ha) and the number of individual amphibious specimens (more than 130,000) to be found there. The information that the compensation measures are to be completed by 2011 suggests that the panels were erected before 2011.

I recall the two other panels, or double panels, I had seen earlier, located before the bridge. These two panels, also without authors, are dated the 11th of June 2022, the day after the Lacoma villagers' gathering, which I mentioned at the beginning of this chapter. Both panels are written in German and Sorbian. The left-hand panel, entitled "Lacoma – a place of protest", tells a story from 1983 to "today". The story starts with the residents' opposition to the destruction of their village and their unwillingness to sell their land. The protest against the SED is detailed here: women from Lacoma went to Berlin to protest. The protest achieved a concession by the government, the collective relocation of villagers to a new location with new homes. Subsequently, a priest and environmental activists continued protests against the village's destruction. In 1991, the first Lacoma festival was held in an effort to revive the village. Protestors prevented the felling of the village oak. Young squatters arrived in 1992, and their presence was legalized in 1993. Subsequently, the village experienced a revival period; village traditions were recovered, and Lacoma hosted a variety of cultural and environmental events. In 1994, a memorial site for the destroyed villages was founded – the white crosses I mentioned earlier (Figure 7.5). Protests continued into the 2000s, for instance, against extra-legal mining activities. By 2022, a part of the earlier village had long been flooded by the new post-mining lake. In passive form, the panel narrates that it was suggested that the lake be named after one of the destroyed villages, but Cottbus town councillors did not support the idea. The right-hand panel focusses on the pond landscape and the Old Hammergraben Canal, framing these as "lost treasures". It refers to several administrative voices from the municipal and regional government that consider these treasures to be of the highest natural, cultural and historical value, noting that the Old Hammergraben Canal was created for industrial purposes around 1550. The panel evokes 24 ponds in a space of 300 ha, and the ecological features between these, like wet and dry lands, which are home to some 170 endangered species, including some "priority species". From 2003, the area was assigned the European Union's nature protection category, Flora-Fauna-Habitat (FFH). The destruction of that habitat is marked as the single known case of the complete destruction of such a habitat. The administrative decision that allowed for the destruction occupied courts around a decade till 2016 and involved the courts recognizing the violation of foundational rights (without specifying which rights). The destruction allowed for the excavation of some 48 million tonnes of lignite, which powered the nearby power plant for two years.

Following the footpath onto the embankment, the horizon widens. I see a lake and beyond it, dry lands. In the far distance, there are trees, and I count 18 wind turbines. A bench provides an effective vantage point; an open hut complements this scenic outpost. I turn my gaze back to the remaining homestead next to the New Hammergraben Canal, but it is almost hidden behind the large tree and the hut. Up here, except for the tree, pre-destruction elements of La#oma appear nearly absent. While I am at the outpost, an old couple briefly visits and comments on the view, "quite beautiful here", and moves on. After a while, fire brigade officers arrive. We started to chat, and I learned about how they use this place for large-scale training exercises. As I walk down, I find three brigade cars in the parking lot. Across the street, on the side of the last homestead that faces the public street, I see a bulletin board with an invitation to the 2022 Lacoma Festival as well as an announcement for a flea market. Leaving the place, behind the bridge, I see a teenager walking across the street towards La#oma but they turn and reach the bus stop, where they unlock their bike and cycle off.

La#oma ont*-politics

Looking through carbon can generate analytical stories of how places *of* and *against* carbon are multiple and generate existential "politics of what". La#oma can be positioned in a broad geographical and political landscape. I find such positioning within geographical maps and of the village in relation to other destroyed villages, both in the Archive as well as at the site of the white crosses. While discussions of the ontological politics of maps are not new (Gerlach 2014), my focus exceeds the politics of maps-as-documents and scrutinizes the relation between the visitor and the wider landscape of villages in which La#oma is located. To tease out the ont*-political character of La#oma, I need to attend to the humans (myself as well as observed or imagined other visitors and users) and their con-figurations, their part in the relational constitution of the village and their concerns with environmental destruction.

La#oma and its configurations with human actors

The crowd of villagers I met in the summer of 2022 included squatters, legalized tenants, dwellers in huts and tents, local business people and anarchists whom I knew from performing direct action. I wonder how human actors are con-figured in the two walkable places I mention here. In the Archive, the visitor relates to these destroyed villages by walking across them. They appear significant as red marks on the floor map. Via the Infosauger, the visitor-as-user looks down onto the map to find and then open the respective site for accessing a village's resources. The village is rendered accessible through the Infosauger's touchscreen navigator, and details of the village are revealed, at least partially, by the user's commands. The user can retrieve information, leave the village and move on to the next. In this form of access, the visitor is con-figured akin to an extracting agent, like the mining industry, moving across the landscape, retrieving resources from the ground and then moving on.

To get to the geographical location of La#oma, the pavement that leads from the bus stop passes by the white crosses. I read these as symbolizing unnecessary death, as mobilized in other activist protests (Holmes and Castañeda 2016). In the green, more or less grassy verge next to the main street and the pavement, the crosses call for attention and ask for mourning. The crosses con-figure a human that stays outside of the landscape of the crosses. The destroyed villages are represented, but the visitor (as an actor) does neither walk through nor over these villages (as with the map in the Archive). A visitor who stays on the path can only look at the crosses from a distance (a couple of metres), and pause to think, remember, mourn or feel agitated. The windmill provides a resource for these reactions, as the windmill might be read as indicative of an alternative to the fossil-fuel-based energy economy, which is available but not employed. At the same time, actors are not necessarily drawn into a critical and mourning relation to the village, as evidenced by the tourists and firemen I mention above.

Actors at both sites experience different con-figurations – at the Archive, actors are con-figured to extract; at the geographical location of Lakoma, visiting actors can experience con-figurations for mourning but also for lake use (as tourists or firefighters); the actors who set up the panels and those who designed/maintain the Archive con-figure others as readers, shaping their imagination and understanding. The actor con-figuration in La#oma is multiple, too. I want to highlight how these two actor con-figurations appear to contrast, even though both require the actor to face both the village identity/identities as well as their destructions. And particularly, I suggest that the storying of La#oma and its land is conflicted: actors are con-figured as accomplices in extracting resources from the land and the village while at the same time they are invited to question, fight and mourn that very extractive mode of relating to it.[13]

Relationally constituting La#oma's naturescultures

My visits to La#oma at the two sites point to two key ways in which the village can be engaged within the present. While above we encountered the visitor and user as subject(s) of inscriptions and infrastructures, I now analyse the village as a subject. It becomes subject to the visiting actor, engaged in recreation, tourism, learning or research. Beyond that, in the village, I find indications of commerce happening at weekends, but also that people dwell in La#oma – recall the last homestead and the bulletin board – marking a space of public interaction and living a dynamic day-to-day life. The fire brigade functional site No. 7 signals the role of the place as part of the rescue and safety infrastructure that surrounds the post-mining lake.

While various practices and forms of life intersect in contemporary La#oma, the village is also enacted through inscriptions. I am fascinated by the contrast in the framing of the relationship between mining and the village. Both the Archive as well as the three-sided panel in La#oma frame the "devastation" of the village as a given necessity. La#oma's existence is placed in the past (which contrasts with the view of the activists who organize the Lacoma festival, which enlivens La#oma in the present). I interpret that necessity as part of the commitment to mine for carbon fuels; however, the necessity seems taken for granted, and it does not need to be explicitly evoked within each situation in which it performs reality. In these two instances, carbon makes the destruction of villages and their landscapes necessary. In contrast, the two panels installed in 2022 frame the destruction of the village and its landscape not only as *not* necessary but also as deeply contested. Fittingly, the three-sided panel relativizes the protest: directly after making it present, the protest is then hidden behind the message that villagers sold their land. That message gets complicated by the 2022 panels, with the clarification that the financial arrangement was not only desired but also protested against. Let me magnify the contrast in the relations of these naturalcultural stories to carbon: one story suggests the natural destiny of carbon to be mined and La#oma to be devoured, while the alternative story insists "it could have been otherwise", carbon did not have to be, and should not have been, mined.

The relation between the village, the landscape and nature appears in different ways across the sites. In the Archive's Infosauger database entry on La#oma, the village appears primarily as a cultural and historical artefact. The ponds appear as material artefacts of human culture. Green exists in the margins, marginal to the explicit account. Apart from coal as the necessitating agent of destruction, non-human entities do not play a significant role in the Archive's representation of the village. In the Archive's imagination, nature is other to culture.

That opposition between nature and culture is presented differently in the three-sided panel. There, the village, the fishery function of the ponds and their human management are considered key to the production of valuable natural space. However, the details of this are not significantly mentioned, and instead, the compensation measures are presented with details of their ecological achievements. This hides the ecological destruction and the brown transformation, and it foregrounds human intervention, making and re-making nature ecologically, and even relocating the River Spree to its "original" river bed. In this three-sided panel, nature is culture; it is (green) infrastructure that is optimized for producing value (from a fishery resource to its ecological functions) and that needs to be maintained, continuously subject to transformations. However, in this version, mining is other to nature; mining substitutes pre-existing valuable natural spaces, and mining requires ecological compensation.

The 2022 two-sided panels narrate a competing story that foregrounds commitments to living with and in the natural environment of the village and the Old Hammergraben as well as the recognition of the canal landscape's conjoined cultural-ecological value. That value is signalled, *inter alia*, by recounting lost natures. The destruction of that landscape is quantified in terms of the years across which the extracted lignite was able to power the nearby power station, which in turn sustained the continuing brown transformation, sustaining unsustainability. In that way, coal, the landscape lost, including the village and the villagers, establish a knot of cultural-natural-historical-industrial relations. In the 2022 panels, nature takes on a nearly postmodern role: it becomes part of a knot of relations in which the human is always involved in varying ways – as observer, protector and user. With Latour (1993), one could say that this simultaneously builds on the modern understanding that nature can be measured across distance and a postmodern understanding in the sense of undermining the distanced knowability of nature by way of placing the human within it (see Latour 1993, 46).

To draw this together: while La#oma was shaped economically, politically and violently, the situated modes through which La#omas now exist also shape La#omas' readers, visitors and otherwise con-figured actors. Contrary to the narrative of the Archive and the three-sided panels, the village, in its multiple existence, exists not only through the past but also through ongoing relating. The multiple facets of the inscriptions demonstrate the contested imagination and meaning of the village and its environments, including disparate natures-cultures, where some neatly separate nature from culture, others figure nature in a way that makes it compatible with the discourse of green infrastructures, and yet others evoke a postmodern knotting of natures and cultures.

Erasing and easing environmental destruction

To detail the politics within the relational production of La#omas, I attend to the performativity of material and embodied relations. I focus on how these achieve not only the invisibilizing – aka erasing – of environmental destruction but also its discursive rendering as self-evident, easing such a destructive transformation.

The Archive, as a designed and constructed space, provides access to a visiting user, who is activated to roam the mining district, manoeuvring the Infosauger across the landscape to draw resources from the ground in order to shape the user's knowledge. The user-knower is activated as an embodied operator of a device that, without that embodied interaction, would be ineffective. As argued above, the user-knower's body and the practice of knowing are materially tied together in this space, not just generating knowledge but also entraining the body to enact resource gathering akin to the mining industry's *modus operandus*. Beyond this recognition, in a Verranian way, we can consider the visiting user handling the Infosauger as being *committed*-qua-handling to a reality of the region that can be walked across and runover. The visitor is complemented with a device that eases the knowing of the region's resources through their extraction. The handling of the Infosauger to get to know the region enacts an ontic politics, a commitment to a reality that is subject to the resource gatherer while seemingly entraining the human subject "collaterally" as an extractor (Law 2012).

In this way, the Archive emerges as an organized set-up con-figured not only to achieve epistemic effects. It is also performative of non-epistemic reality effects as the body is entrained. This is what Barad (2007) refers to as an apparatus, and with this notion I am guided to emphasize how the material and epistemic effects, including their politics, are shaped, even if such effects and politics are not constrained to the Archive but are also discernible, for example, at the village site.

The three-sided panels tie La#oma's brown transformation to the performance of the Spree's relocation as green. This might be criticized as carbon being greenwashed (Lippert 2011b) – for linking mining with the ecological restoration of the river is depicted as if the mining project in itself was not environmentally destructive but generative, even though the mining project's problematic transformative effects are left untouched and hidden behind the green story of the river's relocation. In the three-sided panels' performance, without humans, ecological nature would not exist, and only because of the compensation measures is the river (as it were) restored to a more natural state. I read the panels as communicating the following: mining achieves a green transformation of the river ecosystem. This constitutes an epistemically and politically relevant justification of carbon extraction "after the fact".

With Barad, the humans who perform this restoration work appear as central to enabling the post-La#oma regional reality (the mine and its subsequent more or less utopian state of a lake). These humans matter in this reality, for they shape not only how La#oma is known but also what it is: a sad but, supposedly, transformed green place. In this apparatus of La#oma, the village, its past and its future have been con-figured together with the river that was "naturalized" elsewhere. If we focus on the river, in it, we can sense the ghost of La#oma (beware, below you will encounter other ghosts).

Sensitized by Mol's (2002) take on different material practices at various locations, the Archive's con-figuration of the user emerges as a form of ontological practice, but this also happens in a range of sites in La#oma. Here, I note four quite distinct ways of enacting the village/nature/landscape. As a reader of the different panels encountered, both the three-sided and the double panel, the human joins in reading and narrating specific stories, which epistemically diverge. For instance, we find in one set of panels (the three-sided panels involving "Lakoma and mining") a version of La#oma that is characterized by a centuries-old history, with no significant recent history post-1989. The alternative set of double panels (set up by the activists) presents a recent history with a hopeful vision for environment–human–culture relations, enacted between 1990 and 2007. Contrast these user-as-reader positions with that of the passing-by pedestrian next to the white crosses, who is asked to pause, mourn or get enraged. Insofar as the white crosses assemble the destroyed villages as dead entities, this assemblage temporarily con-figures the passer-by as being with death, with destruction. The ghost of La#oma meets its ghostly family. At the bulletin board in the last remaining homestead, the one-time visitor can act as a reader, but a regular passer-by can also equip themselves and bring devices (like notes and pins), acting as a writer, to interact with the public through inscriptions or otherwise, thus enriching the public life of La#oma and therewith changing what La#oma is, effecting ontological politics.

Across La#oma's different realities, diverse agencies that bring the village into being appear. Sensitized by the continuum between ontic and ontological practice (identified in Raasch and Lippert's (2020) reading of Verran's oeuvre), I illustrate these agencies. In the Archive, a clear set of actors is in control of design and maintenance. The Infosauger does not con-figure the user as a contributor who might contribute with their knowledge of a vanished village. Erasure here takes the form of not allowing a voice[14] – the Archive's knowledge infrastructure does not invite visitors to change the records, which are only accessible via the Infosauger. Similarly in control over what La#oma is are the authors of the various panels. The visiting user is not invited to inscribe anything into La#oma there. The panels provide distinct ontologies of nature – a nature-culture within the 2022 panels versus the marginal appearance of nature as an achievement by the compensation measures within the three-sided panels (from before 2011) versus cultural and historical stories distributed across all the panels. In the Archive, nature as an "other" appears as an ontic effect. There, the authors and designers have effectively con-figured an infrastructure that continually enacts the splitting of naturescultures into culture and nature.

Erasing seemingly critical histories, and silencing natures from mattering in some of these versions of La#oma, eases environmental destruction. This effect is mediated not only by an ontological politics of destroying La#oma, of telling of the mining as achieving a green transformation, but also by the commitments con-figured through stories and their infrastructures implying how humans relate to La#oma. I think of these commitments in terms of ontics

– politics of what is – which comes to the fore, *inter alia*, with the user of the Archive or the user of the three-sided panel who treats that panel as a read-only device, not subverting its presentation. However, this analysis also shows that La#oma's varying technologies of inscriptions are not arbitrarily placed but con-figured to achieve targeted constitutions of the village. These con-figurations are locally organized in the form of apparatuses that con-figure actors and subjects of the village, while that village exists in forms of particular and contested naturescultures.

(Re)imagining La#oma ont*-politically

What does it mean to imagine La#oma ont*-politically? And is it possible to reimagine versions of La#oma ont*-politically? To explore the productivity of these questions, I speculate I use my empirical material for grounding purposeful speculations through which I try to imagine alternative La#omas.

The bulletin board equips the passer-by: as a pin-able board, the passer-by, with minimal resources (pins and paper), can invest and inscribe their own commitments, thereby shaping the place. While such shaping is already invited by the board, the outpost invites me to speculate. The outpost is infrastructured as a recreational space. Lacking critical memorial infrastructure, the outpost's user would have to mobilize significant material equipment to achieve a long-lasting generative shaping of the place; for instance, it would require a massive intervention to set up a new panel that makes the village, the ponds and the Old Hammergraben canal visible and imaginable (say, e.g. a transparent material with relevant drawings and other inscriptions).

However, even when we try to render some of these relations imaginable, the question arises: through which commitments do we infrastructure the work of imagining what was, what is, and what could have been? The Archive's inscriptions render my attention sensitive: what ghosts of extractivist ontologies haunt our remembering of lost villages and their naturescultures? The different sets of panels that represent La#oma indicate the different commitments to what the village, its environment and Lusatia have been, are, could have been or could be. This only starts with contrasting the green transitions embraced by some of the activists with the mining industry's and their partner's commitment to the region as a reservoir of lignite. We can continue at a much more detailed level of observation if we recognize the contrast between the commitments to how La#oma's destruction was destructive (e.g. the two-sided panels) or conversely constitutive of nature conservation (e.g. the three-sided panels). The two competing sets of panels each tell partial stories. I wonder: what might a heritage device, a locally situated knowledge infrastructure that can attune its visitors to the multiple, non-coherent commitments to realities present across the La#omas, look like?[15] I expect we would need quite an amount of resources to conduct the work of accountably (re)imagining La#oma, while also committing to La#oma's conflicted, partial and contradictory pasts, presents and futures. However, without investing in such resources, in the early 2020s, the outpost's gazing subject is con-figured to overlook (ignore) the village, drawn into the lake, a "new nature, perhaps". As it were, during my fieldwork phase, the three-sided panel in La#oma indicated a village life that was committed to a version of village identities supposedly dominant up until the late 1980s. In contrast, the activists, squatters, punks and others I met in the summer of 2022 indicated multiple ontologies of differently colourful commitments, practices of knowing and materially achieving through tents and other construction work and through cultivating relations with a range of species, like the storks mentioned earlier. Some of these latter en-lived relations have been framed explicitly as political, as thoroughly indicated on the two-sided panels.

I reconstruct these experiences and readings of La#oma as permeated by ont*-politics, where what La#oma is and what it becomes is shaped by divergent commitments, saturated by stories of conflict and tension, even including stories of the military and insurrection. Mining lignite was not only not necessary, but it was and continues to be contested. At the same time, within the apparatuses of La#oma and the Archive, an ont*-politics is enacted that achieves La#oma as a desaturated village, "devastated" out of "necessity"; a place void of a natural nature, overwritten with a story of mining carbon that is ascribed with the agency to regenerate nature and even to bring back original nature. In that version of La#oma, its destruction was for the environmental good. My partially pragmatic realist self considers the latter version of La#oma denialist, as it denies and erases the environmental destruction, the 48 million tonnes of lignite that sustained the power plant for two years, and continues to be destructive by way of strengthening calls for the continuation of opencast mining.

Conclusion: the multiplicity of a village entwined with green and brown transitions

I set out by introducing Lakoma as a contested place in which different names of the village do not simply underwrite but primarily resonate with different communities committed to human and/or more-than-human life there. To reference the village, I used the neologically constructed technical term La#oma[5] to bring forward this more-than-one identity of the place consistently throughout the chapter.

Investigating various stories that related La#oma to carbon, I can distil distinct transitions. In the corporate traces, I find the mining of carbon as a necessity, a regional self-evident commitment to sustaining the brown transition in the pre-1990s. It is important to note that since the 2000s, this brown transition was stabilized in these stories by linking mining closely to a supposedly green transition in the form of relocating the river Spree. In that storying, Lakoma's destruction was for the good. That version of a green transition is challenged by activists' vision for a green transition that was invested in sustaining the lakes, lands and naturescultures of Lacoma without environmental destruction. The activist-squatters did temporarily achieve a transition of the nearly evicted village population into a rich and thriving counter-cultural community existing despite mining and against mining. These reconstructions imply competing realities of what La#oma is and was, and how it did and could transition.

To sociologically inform my analytical imaginations of the realities of La#oma, I have mobilized a range of recent STS theoretical resources, developed within post-actor-network theory and new materialism as a toolbox that allows for studying ont*-politics. Ont*-politics is a "politics of what" that is not necessarily theorized and captured by specific epistemic forms. It is rather shaped by the entanglement of what is and how that which is, is imagined, known and/or (practically and materially) committed to, but without any determining of becoming. After introducing the context of my empirical materials and methods, I detailed how La#oma is at two walkable sites – the Archive and Lakoma – a still-existing, geographically and mapped place. I focussed on the reality of La#oma in its relation to various natures, greens and browns, from lignite to ponds and amphibians, and the enactment of said relations through a range of communities and interested parties. In my analysis, I covered the actors con-figured in the two sites, the relational reality of the village and the erasing of destroying natures for easing brown transformations. This led me to (re)imagine La#oma ont*-politically.

Carbon figures in this analysis as agential of mining and energy generation, of destruction of natures as well as the generation of natures, in destroying the village, to the village's revival and resistance. I showed the modalization of mining in the region as natural, rendering La#oma as a place of, and for, carbon mining. I contrasted the radical green transformations of mining plans imagined by the activists (the squatters of La#oma, seeking to prevent mining by turning the village into a place against mining) with the mining industry's seemingly innocent green-blue transformation of the lignite mine into a lake (and renaturalizing the nearby River Spree). This constituted La#oma as a place of post-carbon mining, a place for green-blue transformation and against green transformation (by preferring a brown transformation, mining, over an immediate stop to mining). In these multiple realities of La#oma, carbon haunts each narration and each location differently. Carbon thus not only superficially shapes the landscape through calling for mining (e.g. the mining industry and the Archive), but carbon is also an immaterial presence in the ongoing remembering of La#oma and in the competing enactments of what La#oma was, could have been, is, and might become.

The substantive discussion illustrates the productivity of analysing the politics of reality-making and becoming ont*-politically. I detailed practices and experiences of encountering versions of La#oma that are haunted by commitments to extractivism that are partially legitimized and partially hidden. I problematized how some realities are erased, while others are sustained within competing accounts.

This analysis serves as a ground for speculation: what are the epistemic, imaginative and material resources needed to design a different infrastructure for remembering La#oma, to re-con-figure La#oma-as-apparatus? As a starting point for such work, I suggest that La#oma is a village that exists not only in different places – within the geographical space of Lakoma, but also in a museum removed from the "original" site – in which it is enacted differently, but it also comes into being as a contested and multiple reality. My line of analysis indicates a lost opportunity in the way La#oma's multiplicity is employed at the two sites, through the competing infrastructures of enacting the village and its naturescultures as well as the disparate knowledges of it. There is an opportunity to mobilize these towards fostering discursive engagement with the political sensibilities in remembering La#oma and its relations to carbon across difference. From what I observed, I did not find storying that operated through these conflicting naturescultures. Yet, energy transition discourse in the region could profit from a nuanced and non-singularizing account of the conflicts still enacted in the causation of and suffering resulting from lignite mining.

In short, I suggest that the story of La#oma provides an opportunity to consider how a mining region remembers itself and how it empowers itself to critically re-con-figure and emancipate itself from dominant commitments to ongoing brown transformations. How the village and its environs are told, how its users are con-figured to enact La#oma – these are concerns that matter to what La#oma is and what it becomes. Heritage politics in the region might profit from reconsidering its infrastructure, which actively performs practices of remembering the destroyed villages, including their naturescultures to boost the green transitions needed in the early decades of the 2000s. A strategy attentive to multiplicity seems generative of the required understanding of the village/landscape/natures as thoroughly ont*-politically constituted by subjects, agencies of place and materiality, including humans and their intra-relations that con-figure each other.

Acknowledgements

The research underlying this chapter has been supported by the project Sociocultural Carbon (SOCCAR), funded by the Danish Independent Research Fund, as well as by the Bureau for Troubles of the Museum für Naturkunde Berlin and its research travel backing. It also received institutional support from the Chair of Technoscience Studies at Brandenburg University of Technology. The research process has been deeply shaped by a conversation with Lesley Green. Work on this chapter was enabled infrastructurally by the child care work of my co-parent. The chapter's development is deeply shaped by the patience and detailed engagement with it by my co-editors, Katinka Amalie Schyberg, Andy Lautrup and Steffen Dalsgaard. Anonymous copy-editors smoothened the language and I used ChatGPT (Version 2, October 2023) to review some of the copy-editors' language suggestions to balance general English reading expectations and analytical precision.

Notes

1 Fellow students and I cherished experiences in Lacoma, with its "valuable cultural and natural landscape" (Weber in Lippert 2005). The, at that point potential, destruction of Lakoma meant to us a normative occasion, prompting us to problematize the concept of "environmental management" and its commitments to sustain the exploitation of a world framed in terms of resources.
2 For a complementary analysis that engages with the Wikipedia representation of Lacoma, see Lippert (2018).
3 Note, I use "transition" and "transformation" interchangeably.
4 The Sorbs are one of the four legally recognised minority peoples in Germany (Bundesrepublik Deutschland 2019).
5 I borrow the sign "#" (hash) from programming culture, where it often acts as a placeholder for a missing or variable character. In writing La#oma, the # could be "c" or "k", reflecting the multiple spellings of the village's name across languages and histories. Strictly speaking, I could write "#a#oma" to also account for the Ł in Sorbian spelling or even "#a##ma" to reflect both the consonantal and vowel shifts in different spellings of the village's name. However, for readability, I use just one placeholder. This small disruption not only resists linguistic simplification but also gestures toward the historical erasure of Sorbian presence and identity in the region – part of a broader pattern of suppressing autochthonous and Indigenous communities in (post)colonial struggles (see also note 14). Notably, the village's contested naming also plays out in contemporary struggles, as seen in the activist use of "Lacoma" (see Lippert 2018).
6 Mol (2002) elaborates how the disease atherosclerosis is achieved in different wings of a hospital differently. For instance, in one hospital section, atherosclerosis is enacted as something that can be addressed well by a walking therapy and the body is made to walk to change its atherosclerosis. In another section, atherosclerosis may be located deep within the human body and the patient might be cut open with scalpels to treat atherosclerosis. Thus, the same, atherosclerosis within a body, can be enacted differently, resulting in the analysis of the body as being multiple, rather than many.
7 Barad (2007, 163) discusses an experiment set up by Gerlach and Stern, who aimed for an apparatus that was to serve differentiating silver atoms by orientations of the "electrons orbiting the nucleus." Alas, in their initial set-up of the apparatus, Gerlach could not identify any traces of the silver atoms. However, when Stern, as an embodied-socio-culturally-materially-equipped entity, became part of the apparatus, they could find traces: Stern smoked a cheap cigar, which emitted sulphurous fumes that reacted with the silver atoms, thus effecting a visual trace. Barad suggests that the configuration of Stern-as-smoker was quite typical for the place and time of his kind of work.
8 This matters not only for my analysis, but also for questions of design and intervention, as Verran (2007b) indicates in her discussion of IT infrastructure design for indigenous knowledges (TAMI).
9 Similar to the hash (see note 5), I borrow the "*" (star or asterisk) from programming culture, where it functions as a wildcard, allowing for multiple possible characters or word extensions. In my neologism "ont*", the star signals overlaps and gaps between "ontics", "ontology" and "onto-epistem-ology" – *without assuming a neat theoretical integration*. Instead, "ont*" marks a generative analytical space for thinking about the being and becoming of reality.
10 Note that Barad's (2017) concept of ghosts differs significantly to Gan et al.'s (2017). For the latter, the ghost refers to traces of past human practices.
11 The vantage points of this empirical (auto-)ethnography are my former base at Berlin, my former precarious academic job at Brandenburg University of Technology and my participation in the "Sociocultural Carbon" research project from which this volume, as well as Lautrup's (2022) and Schyberg's (2024) theses, emerged. Above, in the second section of this chapter, I detail my positionality in this research.

12 Neu-Horno means "new" Horno. For the history and politics of that village, see Michel (2005).
13 For a related analysis of conflicting configurations of actors in shaping carbon realities, see Hougaard (this volume).
14 I draw here on Falzetti's (2015) discussion of how settler archives render indigenous peoples absent, and as such erase their presence.
15 This kind of wondering might well continue thinking with Verran's (2007b) work of exploring requirements for infrastructures that allow for noncoherence, see also note 8.

References

Archiv der verschwundenen Orte (AvO). 2010. *Dokumentation bergbaubedingter Umsiedlung*. Forst.
Archiv der verschwundenen Orte (AvO). n.d. *Verschwundene Orte*. Archiv der verschwundenen Orte. Accessed July 21, 2024. https://web.archive.org/web/20240721 152559/https://www.archiv-verschwunde-orte.de/de/verschwundene_orte/verschw undene_orte/70543.
Barad, Karen. 2007. *Meeting the universe halfway: Quantum physics and the entanglement of matter and meaning*. Durham and London: Duke University Press Books.
Barad, Karen. 2017. "Troubling time/s and ecologies of nothingness: re-turning, re-membering, and facing the incalculable." *New Formations* 92 (31): 56–86. https://doi.org/10/gf3cwh.
BfN. 2024. *Steckbriefe Natura 2000 Gebiete in Deutschland: 4152-303 Lakomaer Teiche*. Accessed July 21, 2024. https://web.archive.org/web/20220429193816/https://www.bfn.de/natura-2000-gebiet/lakomaer-teiche
Blühdorn, Ingolfur. 2007. Sustaining the unsustainable: Symbolic politics and the politics of simulation. *Environmental Politics* 16 (2): 251–275. https://doi.org/10/bfn8p5
Braunbeck, Helga G. 2020. "The Past Erased, the Future Stolen: Lignite Extractivism as Germany's Trope for the Anthropocene." *Humanities* 10 (1): 10. https://doi.org/10/nh37
Bundesrepublik Deutschland. 2019. *Fünfter Bericht der Bundesrepublik Deutschland gemäß Artikel 25 Absatz 2 des Rahmenübereinkommens des Europarats zum Schutz nationaler Minderheiten*. https://web.archive.org/web/*/https://www.bmi.bund.de/SharedDocs/downloads/DE/veroeffentlichungen/themen/heimat-integration/minderheiten/5-fuenfter-staatenbericht-rahmenuebereinkommen.pdf?v=4&__blob=publicationFile.
Cunningham, Stuart. 2013. "Wends and the Wende: Modern German Unification (1989–90) and the Sorbs." PhD diss., The University of Manchester (United Kingdom). https://www.research.manchester.ac.uk/portal/files/54545387/FULL_TEXT.PDF.
Falzetti, Ashley Glassburn. 2015. "Archival absence: The burden of history." *Settler Colonial Studies* 5 (2): 128-144. https://doi.org/gpbpgs
Förster, Frank. 1998. *Bergbau-Umsiedler: Erfahrungsberichte aus dem Lausitzer Braunkohlenrevier. 183*. Spisy Serbskeho instituta 17. Bautzen: Domowina.
Gan, Elaine, Anna Tsing, Heather Swanson and Nils Bubandt. 2017. "Haunted landscapes of the anthropocene." In Tsing, Swanson, Gan and Bubandt, G1–G14.
Gerlach, Joe. 2014. "Lines, contours and legends: Coordinates for vernacular mapping." *Progress in Human Geography* 38 (1): 22–39. https://doi.org/f5twsm
Gesing, Friederike, Katrin Amelang, Michael Flitner and Michi Knecht. 2019. "NaturenKulturen-Forschung." In *NaturenKulturen: Denkräume und Werkzeuge für neue politische Ökologien*, edited by Friederike Gesing, Michi Knecht, Michael Flitner and Katrin Amelang, 7–50. Edition Kulturwissenschaft, Band 146. Bielefeld: transcript.
Hagemann, Jenny. 2022. "Vom wendischen Notschrei und von sorbischen Kultur-Hochburgen-Nutzungen und Aneignungen kulturellen Erbes im wendländischen

und Lausitzer Protest." *LĚTOPIS. Zeitschrift für sorbische Sprache, Geschichte und Kultur. Časopis za rěč, stawizny a kulturu Łužiskich Serbow* no. 2, 6–23.

Holmes, Seth M. and Heide Castañeda. 2016. "Representing the 'European refugee crisis' in Germany and beyond: Deservingness and difference, life and death." *American Ethnologist* 43 (1): 12–24. https://doi.org/10/gcnzpb

Latour, Bruno. 1993. *We Have Never Been Modern*. New York, London: Harvester Wheatsheaf.

Lautrup, Andy. 2022. *Generation Carbon - Goodness, Loss and Youth Climate Activism in Norway's Oil Capital*. PhD Thesis. IT University of Copenhagen. https://pure.itu.dk/en/publications/generation-carbon-loss-goodness-and-youth-climate-activism-in-nor.

Law, John. 2008. "On sociology and STS." *The Sociological Review* 56 (4): 623–49. https://doi.org/bzxnmh

Law, John. 2012. "Collateral realities." Chap. 8 in *The Politics of Knowledge*, edited by Patrick Baert and Fernando Domínguez Rubio, 156–78. London and New York: Routledge. https://doi.org/10/nh36

Lippert, Ingmar. 2005. *Map-Making for ERM-Studies*. Bachelor thesis, Cottbus. https://d-nb.info/1114664235/34

Lippert, Ingmar. 2011a. "Extended Carbon Cognition as a Machine." Revised version (29th March 2013). Archived at https://nbn-resolving.org/urn:nbn:de:0168-ssoar-60504-9, *Computational Culture* (1). Accessed December 5, 2011.

Lippert, Ingmar. 2011b. "Greenwashing." In *Encyclopedia of Green Culture*, edited by Kevin Wehr, 421–30. Sage Publications. https://doi.org/10/r6f.

Lippert, Ingmar. 2015. "Environment as Datascape: Enacting Emission Realities in Corporate Carbon Accounting." *Geoforum* 66: 126–35. https://doi.org/10/wx8.

Lippert, Ingmar. 2018. "'Earth ... without us', earthlessness and autochthoneity: discursive presences and absences in negotiating and contesting coal mining in Germany." Paper presented in workshop *"Equivocal (anthropo)cenes: indigenous ontologies and the ethics of geo-climatic disruption"* at "Universidad Católica de Chile", 9 November 2018. https://doi.org/10/dw35.

Lippert, Ingmar. 2020. "Ont* – Data Politics – a Theoretical Sociology of Verran's Ontics, Mol's Ontologies and Barad's Apparatuses." Paper presented in session *"Hegemony, counter-hegemony and ontological politics I: Theory, Training and Design."* EASST/4S 2020, Prague/virtual. https://web.archive.org/web/20240128173937/https://www-docs.b-tu.de/fg-technikwissenschaft/public/ontics-talk-revised.mp4.

Lohmann, Larry. 2005. "Marketing and making carbon dumps: Commodification, calculation and counterfactuals in climate change mitigation." *Science as Culture* 14 (3): 203–35. https://doi.org/10/bscgg7.

Mårald, Erland and Janina Priebe. 2021. "Sustainability Metamorphosis: An Inconvenient Change." *Nature and Culture* 16 (2): 1–12. https://doi.org/10/nh35.

Michel, Jeffrey H. 2005. *Status and impacts of the German lignite industry*. Vol. 18. Göteborg: Swedish NGO Secretariat of Acid Rain. Accessed January 3, 2024. https://web.archive.org/web/20240103174158/https://airclim.org/sites/default/files/documents/APC18SE.pdf.

Mills, C. Wright. 1959. *The Sociological Imagination*. Oxford: Oxford University Press.

Mol, Annemarie. 1999. "Ontological politics. A word and some questions." In *Actor Network Theory and After*, edited by John Law and John Hassard, 74–89. The Sociological Review. Oxford, Malden: Blackwell.

Mol, Annemarie. 2002. *The Body Multiple: Ontology in Medical Practice*. Durham, N. Ca., and London: Duke University Press.

Morini, Ryan S. 2019. "'What are we doing to these shoshone people?': The ontological politics of a shoshone grinding stone." *American Anthropologist* 121 (3): 628–40. https://doi.org/10/grh66b.

Müller, Reinhard, Lars Hendrich and Jörg Schönfelder. 2015. "Bemerkenswerte Makrozoobenthosfunde in einem renaturierten Abschnitt der mittleren Spree mit

Erstnachweis von Leuctra geniculata (Stephens, 1836) (Plecoptera, Leuctridae) für Brandenburg." *Lauterbornia* (79).
Ninan, Anup Sam. 2011. "Outsourcing Pollution: Clean Development Mechanism (CDM) as Ecological Modernisation." Chap. 21 in *Implementing Environmental and Resource Management*, edited by Michael Schmidt, Vincent Onyango and Dmytro Palekhov, 263–82. Heidelberg: Springer. https://doi.org/10/d8r7zn.
Norgaard, Kari Marie. 2018. "The sociological imagination in a time of climate change." *Global and Planetary Change* 163: 171–76. https://doi.org/10/gjpd29.
O'Reilly, Jessica, Cindy Isenhour, Pamela McElwee and Ben Orlove. 2020. "Climate change: Expanding anthropological possibilities." *Annual Review of Anthropology* 49 (1): 13–29. https://doi.org/10/ghg7pn.
Pflug, Wolfram. 1998. *Braunkohlentagebau und Rekultivierung: Landschaftsökologie – Folgenutzung – Naturschutz*. Springer.
Raasch, Josefine and Ingmar Lippert. 2020. "Pioneer 'Helen Verran'." In *The SAGE Encyclopedia of Research Methods*, edited by Paul Atkinson, Sara Delamont, Alexandru Cernat, Joseph W. Sakshaug and Richard A. Williams. Sage Publications. https://doi.org/10/fft6.
Reaves, John Andrew. 1996. "The Development of an Ecologically Critical Sorbian Literature as a Consequence of the German Democratic Republic's Dependence on Soft Coal as an Energy Source." PhD Thesis, University of Wisconsin at Madison.
Schleich, Joachim, Wolfgang Eichhammer, Ulla Boede, Frank Gagelmann, Eberhard Jochem, Barbara Schlomann and Hans-Joachim Ziesing. 2001. "Greenhouse gas reductions in Germany—lucky strike or hard work?" *Climate Policy* 1 (3): 363–80. https://doi.org/10/ddvctq.
Schyberg, Katinka Amalie. 2024. 'Church and Climate Change in Counterpoint: An Ethnography of Environmental Engagements within the Danish People's Church'. PhD thesis, IT University of Copenhagen. https://pure.itu.dk/en/publications/church-and-climate-change-in-counterpoint-an-ethnography-of-envir
Sioh, Maureen. 1998. "Authorizing the malaysian rainforest: Configuring space, contesting claims and conquering imaginaries." *Ecumene* 5 (2): 144–66. https://doi.org/10/bd6gtp.
van der Sluijs, Jeroen, Josée van Eijndhoven, Simon Shackley and Brian Wynne. 1998. "Anchoring devices in science for policy: The case of consensus around climate sensitivity." *Social Studies of Science* 28 (2): 291–323. https://doi.org/10/fmppc2.
Soch, Konstanze, ed. 2020. *Stasi in Brandenburg: Die DDR-Geheimpolizei in den Bezirken Cottbus, Frankfurt (Oder) und Potsdam*. Der Bundesbeauftragte für die Unterlagen des Staatssicherheitsdienstes der ehemaligen Deutschen Demokratischen Republik. urn:nbn:de:0292-97839465720537.
Suchman, Lucy. 2012. "Configuration." In *Inventive methods: The happening of the social*, edited by Celia Lury and Nina Wakeford, 48–60. Oxon and New York: Routledge.
Swanson, Heather, Anna Tsing, Nils Bubandt and Elaine Gan. 2017. "Bodies tumbled into bodies." In Tsing, Swanson, Gan and Bubandt, M1–M12.
Tsing, Anna Lowenhaupt, Heather Anne Swanson, Elaine Gan and Nils Bubandt, eds. 2017. *Arts of living on a damaged planet: Ghosts of the Anthropocene*. Minneapolis: University of Minnesota Press.
Verran, Helen. 1998. "Re-Imagining land ownership in Australia." *Postcolonial Studies* 1 (2): 237–54. https://doi.org/10/bnmsfx.
Verran, Helen. 2001. *Science and an African Logic*. Chicago and London: University of Chicago Press.
Verran, Helen. 2007a. "Metaphysics and learning." *Learning Inquiry* 1 (1): 31–39. https://doi.org/d27n5s.
Verran, Helen. 2007b. "The educational value of explicit non-coherence." Chap. 6 in *Education and technology: Critical perspectives, possible futures*, edited by David W. Kritt and Lucien T. Winegar, 101–24. Lanham, Maryland: Lexington Books.

Verran, Helen. 2012. "The changing lives of measures and values: from centre stage in the fading 'disciplinary' society to pervasive background instrument in the emergent 'control' society." *The Sociological Review* 59: 60–72. https://doi.org/10/sh8.

Weber, Gabriel and Ignazio Cabras. 2017. "The transition of Germany's energy production, green economy, low-carbon economy, socio-environmental conflicts, and equitable society." *Journal of Cleaner Production* 167: 1222–31. https://doi.org/10/gcgpz5.

8 Scaling the world through carbon

Discursive decoupling, unity and climate responsibility in Stavanger, Norway

Andy Lautrup

Introduction

"Oil is not the problem. Oil is a wonderful product," an oil industry representative said eagerly during a debate about the future of Norwegian oil production I attended in Norway's oil capital, Stavanger, in late 2019. He continued: "The problem is that when oil is combusted, it emits CO_2." The debate took place in a large university auditorium, where two young climate activists heatedly discussed with two older representatives from the oil and gas industry. One of the young climate activists argued that Norway has a historical responsibility to cut down on its oil production. "We have become so rich, and we have let out so much CO_2," he emphasized as the reason for this historical responsibility. Pushing back against the proposition of Norway's historical responsibility to end oil production, the oil and gas representative reasoned that Norway should rather be the last to do so. To support his argument, he stated that the wealth the oil generates is distributed fairly in Norway, which contrasts with countries in the Global South where oil production only benefits a small elite. Additionally, he proposed that Norwegian oil is in fact good for the climate because the production of Norwegian oil has a low carbon footprint.[1] His statements were met with exasperated gestures from the two young climate activists.

This small scene from the debate in Stavanger aptly portrays the core of what I engage with in this chapter: namely, how activists and the industry use carbon to scale the world in ways where the national and the global stand in specific proportions to one another, resulting in different understandings of what it is good and right for an oil-producing nation to do amidst the escalating climate crisis. Scaling the world through carbon, I argue, enables both discounting Norway's climate responsibility and expanding it, depending on how the relationship between the national and the global is enacted.

Throughout the chapter, I show how both activists and the industry are oriented, albeit in different ways, towards doing good as a central cultural commitment in Norway. For the activists, doing good is part of a commitment to global justice. They insist that Norway, and other actors who reap the economic benefit of carbon-emitting activities, has a historical responsibility that

DOI: 10.4324/9781003478669-9

obliges these actors to drastically lower the carbon emissions from the activities their affluence stems from. For the industry, doing good is tied up with the intimate connections between oil and national welfare and the contested claim that Norwegian oil is good for the climate.

Empirically, this chapter is based on nine months of in-person and online ethnographic fieldwork with young climate activists and their local community in Stavanger, which I carried out between 2019 and 2022 as part of my PhD research (on the latter see Lautrup 2022).[2] I participated in the activists' meetings and protests and interviewed them alone, in groups and together with their parents – most of whom work in the local oil industry. I further interviewed people working in the local oil and gas sector and participated in local events related to the oil industry. The young activists I met in Stavanger have grown up marked by the increasing severity of the climate crisis and are currently negotiating how to relate to oil through its carbon-emitting qualities in a context where oil is largely understood as a local and national source of good. Through following the activists' contestations with the local oil industry – with whom they are intimately connected through ties of family and friendship – I became interested in what different scalar conceptions mean for how activists and the industry ascribe responsibility for the climate impact of Norwegian oil extraction and combustion.

Theoretically, I engage with scholarship that views scales not as descriptive, but as cultural and ontologically real claims about the world (Tsing 2000; Simons, Lis, and Lippert 2014). I draw on a science and technology studies (STS)-inspired understanding of ontology: that ontology is dynamic and comes into being through the practices and actions of human and non-human actors (see, e.g. A. Mol 2002; Law and A. Mol 2002; Gad, Bruun Jensen, and Ross Winthereik 2013). Using scale as a framework to analyse industry claims about the separateness between oil and emissions – exemplified in the debater's above-cited statement that the problem is that oil emits carbon upon combustion – I show the generativity of understanding these industry claims as a discursive decoupling of oil and emissions. Discursive decoupling, I suggest, both rests on and enacts an ontological separateness between oil and emissions that enables bracketing out the global from the national when it comes to claiming responsibility for the climate impact of Norwegian oil. Conversely, the young climate activists refuse the industry's ontological separateness and insist on the unity and inseparability of emissions and their source. Hereby, they also refuse the industry's discounting of responsibility and insist on Norway and Stavanger as places from where the global climate crisis can be acted upon locally by phasing out oil and gas production.

Contextualizing Norwegian oil production

Though mostly talked about simply as oil, Norway extracts fossil fuels in the form of oil and gas, both from offshore fields in the North Sea. Since most of Norway's domestic energy use is covered by hydropower, oil and gas are mainly an export business. Oil has a rather recent history in Norway. The first commercial oil field was discovered off the southwestern coast of Norway in 1969, and Stavanger became the official epicentre for the blooming Norwegian oil industry shortly thereafter. This story has been retold in a range of TV series, most prominently the Netflix success State of Happiness (Lykkeland in Norwegian).

Up until the discovery of oil in 1969, Norway was the poorest of the Scandinavian countries, largely because of the country's quite recent history of independence. Until 1905, Norway had been subject to rule by its neighbouring countries, Denmark and Sweden, for several centuries. Through most of the 20th century, Norway was the poorest and least industrialized of the three countries. Venturing into oil production became a financial turning point as the booming oil industry put Norway in a position of being more well-off than its past colonizers.

Drawing on Norway's previous experience with hydropower, Norwegian politicians and civil servants managed to put in place a system with a high degree of national ownership and mechanisms ensuring that income from the oil industry would be channelled back into Norwegian society in the form of wealth and jobs (McNeish and Logan 2012, 2–3; Ryggvik 2010). These mechanisms include a Government Pension Fund, often dubbed the Oil Fund, where the revenue from taxes on oil – as high as 78% – is invested. The fund provides minor inputs to the Norwegian national economy but is largely an independent investment fund with a recent focus on being "a global leader in managing the financial risks and opportunities arising from climate change" (Norges Bank Investment Management 2022). In parallel with the fund's turn towards financial climate leadership, the Norwegian state sustains continuous exploration for new oil increasingly deep into ecologically vulnerable Arctic areas. National Norwegian climate targets centre on reducing national carbon emissions while discounting emissions from the combustion of exported Norwegian oil and gas abroad. Since carbon emissions from oil extraction count as national emissions, the industry focusses on extracting oil in the least carbon-intensive way, even though the vast majority of emissions occur upon combustion.

The present-day focus on reducing emissions from extracting oil is paired with the peculiar claim that Norway produces so-called clean oil. The idea of oil as "clean," as I encountered it among people affiliated with the oil industry in Stavanger, relies on two premises. First, that much oil in the Norwegian North Sea consists of so-called sweet crude oil, whose petrochemical composition makes it less carbon intensive to refine into other petrochemical products. Second, through ongoing efforts to electrify oil platforms by drawing power from onshore hydropower or offshore wind, which can replace turbines that run on gas from the extraction process (Equinor, n.d.). This gas can then be exported, resulting in emissions that do not count against Norway's domestic climate target, because the combustion and the resulting emissions are located in the importing country.

The claim of "clean oil" can be understood as part of a larger narrative of Norway as an exceptional or benign oil producer. Oil-exporting nations have a reputation for being troubled by the exploitative practices of transnational companies, political unrest and authoritarian rule – the so-called resource curse (Behrends, Reyna and Schlee 2011; see Appel 2019 for a critical perspective). However, the Norwegian oil experience has gained the nickname the "oil fairy tale," sometimes also referred to as the "Norwegian model" (McNeish and Logan 2012; Ryggvik 2010), highlighting Norway as both a petrostate and a democratic welfare state with the ability to environmentally regulate and redistribute oil wealth nationally. Though this narrative has been challenged by research into Norway's oil history (see, e.g. Ryggvik and Kristoffersen 2015), it is still culturally prevalent, particularly in Stavanger. When highlighting the superior quality of Norwegian oil, there is often a slippage between cleanliness in terms of what is allegedly a low-carbon production and the social redistribution of oil wealth through the democratic welfare state.

As the oil capital of Norway, Stavanger is a complicated place for climate activism. Most activists' have friends and family who work in the oil industry or its vast supply industry. This is not surprising, since almost 30% of all employment in the region of Rogaland – to which Stavanger belongs – is related to the oil sector (Hernes, Erraia and Fjose 2021). Oil is largely understood as the main driver of positive change in Stavanger. With the oil industry, the inhabitants increasingly went from tending to small farms, fishing a declining stock of fish or working at the local canning factory to flying to work on the platforms in a helicopter or working in the construction of new oil platforms (Gjerde 2002, 55–56; Berge 2014, 14). These new possibilities and a rise in income came to characterize life in Stavanger over the next few decades and continue to frame local conversations about what is at stake in ending Norway's oil production.

Scale as an object of analysis

When I initially noticed the national and the global at play in controversies over climate responsibility in relation to Norwegian oil and gas, I thought of them mainly as descriptive scales. However, the anthropologist Anna Tsing's propositions for how to work with "scale as an object of analysis" (Tsing 2000, 347) shifted my thinking about what scale is and does. According to Tsing, scale is ideological because scaling entails "cultural claims about locality, regionality, and globality," and she encourages researchers to track "rhetorics of scale as well as contests over what will count as relevant scales" (Tsing 2000, 347). Such rhetorics and contestations are part of what Tsing calls scale-making: the "social and material processes and cultural commitments" through which "localities or globalities come, tentatively, into being" (Tsing 2000, 348). Crucially, processes of scale-making are ideological because scale as a framework for understanding the social world defines the parameters for which aspects of social life seem deserving of attention, how they can be measured and to what they can be compared (Gal 2016, 91). Scale, then, is intimately tied to how people perceive relations in the world. What stands in relation to what and in which ways? Which relations matter? When it comes to carbon, scale as an object of analysis assists me in teasing out activist and industry perspectives on what other human and non-human actors and networks carbon stands in relation to and what that means for how they frame climate responsibility in relation to Norwegian oil production.

Taking the lead from Tsing, I approach the different scalar rhetorics at play among oil industry and climate activists in Stavanger not as descriptive, but as perceptions that are continuously produced, stabilized and destabilized with consequences for how people make sense of their reality anchored in the ideological underpinnings of scale. Put differently, scale-making produces a certain ordering of relations between entities as given and unquestionable, and this ordering is intimately tied to a particular world view (cf. Strathern 2020). In their work about scale-making in relation to environmental markets, the sociologist Arno Simons et al. (2014) aptly capture this dual aspect of scale through what they call the "the political duality of scale making," which refers to the "material-semiotic practices of producing and using scales as ontologically real ordering devices" (Simons, Lis, and Lippert 2014, 632). Thinking about scales as both ideological and ontologically real ordering devices connects scale making to questions of power and about who benefits from the stabilization of a particular scalar perspective.

The example at the beginning of this chapter illustrates a general empirical pattern to understand how climate activists and industry people in Stavanger, respectively, perceive the national and the global, and how they use references to carbon to scale the world in particular ways. People affiliated with the oil industry mainly referred to the national level as their point of reference, yet also (more or less explicitly) compared Norway to other oil-producing nations when arguing their extraction is "clean," and they emphasized how oil extraction has been an intricate part of building and sustaining the Norwegian welfare state. While the industry largely invokes oil as a force of good for the national community, the young climate activists see Norwegian oil and gas as a globally destructive force because the carbon emissions from their production and combustion accelerate the global climate crisis. However, the picture is more complicated because, as I lay out in the following two sections, the global also matters to the industry, and the national also matters to the activists.

Industry perspectives: the common good of the national community

Tonje and I sat in her garden in one of Stavanger's residential areas. The late summer sun covered the city in a warm light, and we took refuge in a shady corner. A mutual friend had introduced me to Tonje. "Some say that is the worst kind of work, looking for new oil," my friend told me when discussing Tonje's work, alluding to a common perception that sustaining the oil industry is okay, while it is morally questionable to actively expand it given the increasing severity of the climate crisis. As we spoke in her garden, Tonje explained that she is proud that her work contributes to the Norwegian welfare state. According to Tonje, the oil industry sustains the welfare state, which ensures that Norwegian society has a uniquely high level of equality and a generally high standard of living, qualities she is proud of contributing to and wishes to pass on to her children. Sustaining and expanding the Norwegian oil industry is, for Tonje, integral to preserving the qualities of a society she is part of.

The intimate connection between oil and the welfare state that Tonje articulates is a well-established national and political narrative. Following the 40th anniversary of the discovery of the first big commercial oilfield in Norway, an article in the Norwegian newspaper Aftenposten, published in 2012, underlined this connection. Alluding to the fact that the discovery was made on 23 December 1969, the headline of the article reads: "The Christmas present that became 13 million day-care spots" (Lewis 2012, my translation). The article elaborates that since its discovery, the oilfield has produced oil revenue worth 1791 billion Norwegian Kroner (around 167 billion USD), which the article equates to the cost of 13 million welfare-sponsored day-care spots. The article is a telling example of the broader deep-seated cultural connections between oil and welfare and a pride in the Norwegian model's ability to nationally redistribute oil wealth through the welfare state. In Stavanger, this narrative became salient when the town celebrated its 50-year anniversary as Norway's oil capital on 14 June 2022, and a series of opinion pieces were published in local and national newspapers. "June 14, 1972: 50 Years With the World's Best Oil Model?" is the headline of one such piece, written by two leading figures at the Norwegian Petroleum Museum located in Stavanger. The authors write that "no other country in the world has harvested as much resource rent from its oil and gas resources – on behalf of the national community" (Krogh and Lindberg 2022, my translation). In another opinion piece titled "The Oil History Builds the Future," the Director General of the Petroleum Directorate writes that the directorate "has contributed to making the Norwegian oil industry as good as possible – maybe the best in the world – for the common good of the national community" (Sølvberg 2022, my translation). These pieces reflect a common understanding I came across repeatedly during my time in Stavanger: not only has the discovery of oil been a great stroke of luck for Norway and Stavanger in particular, but Norwegians have also shown a particular skill in their governance of the oil, which has enabled it to benefit the entire national community. A community whose democratic welfare state values and oil wealth should be preserved for future generations.

While aware and concerned about global climate change, most industry people did not think Norway was in a position to do much about the climate crisis, least of all through ending its oil and gas production. "We're too small to make a difference," people often said with references to numbers from The Norwegian Petroleum Directorate stating that Norwegian oil makes up approximately two percent of the global demand (The Norwegian Petroleum

Directorate and The Norwegian Ministry of Petroleum and Energy, n.d.; see also Norgaard 2011 for an analysis of similar narratives). Their reasoning was that sustaining the Norwegian oil industry makes a huge difference for Norway, but ending it will make little to no difference globally. This reasoning was often combined with comparisons between Norway and other oil states, which emphasized how Norway provides good working conditions, environmental regulation and social redistribution of the oil wealth through the welfare state, making Norwegian oil production preferable to oil production elsewhere. I will return to such comparisons in more detail towards the end of the chapter. However, for some in the industry, like the debater from the opening vignette, Norway is in a position to make a global difference by *continuing* its oil production. This argument draws on the aforementioned claim that Norwegian oil is "clean," meaning that it has an allegedly low carbon intensity. And, the logic goes, since the world will need at least some oil for quite a long time to come, from a climate perspective it is better if that oil is Norwegian. Yet, this claim is also contested internally among people in the industry, where many also dismiss such claims, stating variations of what one of my industry interlocutors described as: "Norwegian oil is good for Norway."

Activist perspectives: taking the global lead

"Do you think that people are just like," – Frida's mother stopped talking and put her hands over her eyes, and then continued in a mocking voice – "achhh I can't be bothered to care." "Is it like that?" she asked. It was a sunny afternoon in early autumn, and I was sitting with the 17-year-old climate activist Frida and her mother in their home in one of Stavanger's residential areas. We sat in the garden and discussed Frida's activism and her mother's job in consultancy in the oil industry. Frida sighed and said with a serious look on her face that she thinks some people feel like they can't be bothered or that they don't have the energy. But mainly she thinks people think, "I don't want us to be the ones who sacrifice ourselves, why should we be the ones to stop producing oil, the other countries will just take over?" Frida pointed out that one of the main arguments for continuing Norwegian oil production is that Norway has "the cleanest oil." Frida emphasized that she does not think oil can be clean, and while she agreed that oil can be produced in less carbon-intensive ways, ultimately, "oil is oil," and what matters from a climate perspective is the massive carbon emissions that occur when oil is combusted. "It makes me so mad that people just close their eyes," Frida said, "because I wish that I could also just close my eyes and think oh well, we'll all die, I hope I can live a nice life before it hits me." She sat still for a while, looking at the table in front of her.

I first met Frida two years before, when she participated in the debate that I describe in the opening of this chapter. Clearly, the youngest of the debaters, her strong presence and energetic argumentation made a lasting impression on me. Frida delivered her points forcefully and shook her head in frustrated disbelief at the counterarguments she did not agree with. What struck me about

Frida when I first met her was the force of her sense that something is not how it is supposed to be. For Frida and her fellow young activists, the common good of the national community is not sufficient to justify continued oil extraction. The undertone of their activities ring with the message that because of the climate crisis, it is not possible for Norway to continue to act in the same way as it has until now. They emphasize Norway as both having a historical responsibility for ending its oil and gas production and being in a particularly good position to do so because of its oil wealth. For the activists, Norway's impact on the climate is so big that it would make a global difference if oil production ended. In making this argument, they often referred to numbers from the research, communication and advocacy organization Oil Change International, stating that Norway is the world's seventh largest exporter of carbon emissions because of its oil exports (Oil Change International 2017). Further, the activists think that Norway could take the lead and show that it is possible to phase out oil and gas production without a financial collapse, something they think could inspire other oil nations to do the same.

Discursive decoupling versus unity of oil and emissions

I now turn to an analysis of how the activist and industry perspectives I laid out in the previous two sections rest on different perceptions of how oil and carbon stand in relation to one another. Or put differently, they use scales as ontologically real ordering devices (Simons, Lis and Lippert 2014), enacting oil and carbon as either separable or inseparable with consequences for how they narrate and attempt to frame Norway's climate responsibility.

Recall the oil industry representative who insisted that the problem is not oil, but the unfortunate circumstance that oil emits carbon when it is combusted. Though not often put as bluntly, such a separation between oil and the carbon emissions resulting from its production and combustion is foundational to how people who work in the Norwegian oil industry justify continued Norwegian oil production – especially, as we have seen, through the framework of the national welfare state. To unpack the implications of this way of understanding the relationship between carbon emissions and their source, I begin by identifying the two foundational premises this understanding rests on. First, that oil can be accounted for in terms of carbon emissions. Second, that carbon emissions from oil production and combustion can be dealt with independently from these activities.

The separation of carbon emissions from their sources is aligned with the current global climate governance schemes grounded in the Kyoto Protocol and the Clean Development Mechanism's (CDM) accounting and trading schemes (see, e.g. Newell 2012; Boyd, Boykoff, and Newell 2011). The protocol established that so-called "developed countries" were legally bound to account for and reduce carbon emissions in their national territory. The CDM's carbon trading scheme put in place the basic infrastructure through which these countries could fund carbon reduction projects in the Global South and claim these reductions as part of their national reduction targets. Such offsetting of national emissions through emission reduction projects in the Global South is integral to Norway's national emission reduction scheme and is currently accompanied by promises to develop technology that can capture and store carbon on a large industrial scale.[3]

The anthropologist Hannah Knox (2015) points out how the Kyoto Protocol and the CDM fundamentally changed the technopolitics of oil by making oil conceivable as carbon and thus, by extension, comparable to other carbon-absorbing or -emitting entities (Knox 2015, 314; see also Dalsgaard 2013). For example, following this logic, the carbon contained in oil can be compared to the carbon contained in a forest. However, as Knox crucially points out, making oil equivalent to a forest through carbon accounting is not a question of rendering oil and forests the same. On the contrary, Knox argues, this equivalence is put in place to make other entities, the forest, for example, stand in for the oil in order to "*avoid* the scenario where fossil fuels might have to be conceived as carbon sinks, with the associated risk that this might draw forth regulatory mechanisms that would prevent or slow down the rate of extraction of fossil fuels" (Knox 2015, 314, original emphasis). Conceiving oil as carbon thus enables the separation of oil from its associated carbon emissions so these can be handled separately while leaving oil production itself untouched.

Similarly, ideas about green growth rest on a premise of the so-called decoupling of economic growth and carbon emissions, that is, the hope that the economy can continue to grow without carbon emissions following suit. In economic terms, decoupling refers to a separation between growth and carbon emissions. Decoupling in the economic sense has been discussed and critiqued for decades, for example, in relation to ecological modernization theory and debates over the (im)possibility of sustainable capitalism (see, e.g. A. P. J. Mol and Sonnenfeld 2000 and Foster 1992 for a critical perspective). More recently, critics have pointed out that decoupling presumes tech development as the driving force of decarbonization, thereby seeking to avoid substantial changes to neither the economic system nor individual lifestyles (see, e.g. Dalsgaard 2022). I borrow the term decoupling, but apply it slightly differently, namely with a focus on an industry-driven *discursive* decoupling. Discursive decoupling refers to the way the oil industry discursively separates oil and carbon emissions and how this enables preservation of the status quo while resources are channelled into dealing with carbon as a general entity whose concrete, situated source of emission has little importance. By scaling carbon's relations in the world in this way, carbon can be offset through trading schemes, or it can be dealt with through the technological development of carbon capture and storage projects, thus enabling continued oil production.

Discursively decoupling oil from its carbon emissions rests on accounting for carbon, which was introduced by the CDM. This in turn enables separating national oil from the carbon emissions that its combustion adds to the common global atmosphere. Such accounting enables the industry to locate responsibility outside of the industry and the nation by scaling emissions as a global concern that can be dealt with more efficiently and cheaply through, for example, offsetting practices located in other sectors and parts of the world. Hereby, the national benefits of oil can be preserved without assuming the troubling global responsibility for emissions from exported Norwegian oil that occur abroad. As pointed out by the STS scholar Ingmar Lippert (2011), such industry narratives not only dismiss the relevance of situated emissions but also contribute to a discourse of emissions that enable a universe of references to carbon that allows for business as usual (Lippert 2011). As such, discursive decoupling can be understood as an instance of what the political scientist Ingolfur Blühdorn (2007) calls "sustaining the unsustainable," that is, politics and practices aimed at creating symbolic rather than substantial ecological change (Blühdorn 2007, 266).

By analysing these industry practices as discursive decoupling, I draw attention to how the oil industry cuts up the world in a particular way. This requires a comment on the term "couple," which implies that two in-principle independent entities exist and are, if only temporarily, bound to each other. By using the term decoupling, I do not intend to affirm and accept an already established ontology of emissions as disentangled from their source. On the contrary, I aim to shed light on how the industry conjures up a world in which emissions and their sources are in principle independent, a world in which there is a couple that can be decoupled. And crucially, cutting up the relational network of oil in this way enables claims that carbon emissions from Norwegian oil abroad are not the responsibility of the Norwegian oil industry.

It is exactly the discursive decoupling of oil from carbon emissions, and of the nation-state from the global, that is challenged by the young climate activists' call for assuming a historical responsibility grounded in Norway's economic gains from the carbon-emission-intensive oil industry. Insisting that emissions are inseparable from their source, they locate responsibility at that source and the financial gains it has enabled.

The young climate activists have grown up with the national narrative of how oil benefits both their local and the national community. However, for the young activists, wealth and welfare have little meaning without a liveable climate. They invoke the global climate rather than the national community in their assessment of responsibility for the climate impact of Norwegian oil production. When the young activist debater cited at the beginning of this chapter said, "we have become so rich, and we have let out so much CO_2," he located the nation-state in a global perspective, thereby also insisting on the nation-state as an actor with the ability and obligation to take responsibility.

Crucially, in the activists' accounts, it matters that carbon emissions are specific and localized. They are not just any emissions – they stem from and cannot be separated from the Norwegian oil that Norway gains economic wealth from. And though the oil industry might promote "clean" Norwegian oil, the activists point out that this does not change the fact that vast amounts of carbon is emitted when the oil is combusted. When Frida insists that for her "oil is oil," and oil results in massive carbon emissions when combusted, I understand Frida as also insisting on the unity of oil, with its capacity to be chemically transformed via combustion into emissions. By insisting on unity rather than separation, or on oneness rather than (de)coupledness (where both decoupledness and coupledness share an ontology of two distinct entities – oil and emissions), Frida and her fellow activists evoke the global in a way that enables them to act nationally through the insistence that oil and carbon emissions are intrinsically linked. The activists thus not only insist on the importance of the origins of carbon emissions from a global justice perspective; they also refuse the oil industry ontology in which oil and its emissions *can* be separated.

The activists are not alone in insisting that it matters where emissions originate. They are aligned both with the wider youth climate movement and global political claims for climate justice that push to counter the unequal distribution of climate-related harm. Specifically, high-emitting countries and communities like Norway and Stavanger are often far removed from the geographical locations where climate change presently has the most violent and destructive consequences.

Refusing the industry's discursive decoupling and insisting on the unity and inseparability between oil and carbon emissions has far-reaching consequences for how the activists perceive Stavanger and Norway's oil-dependent status quo: the production and combustion of oil is what needs change, not how carbon is handled. Thereby, they place Norway in a position where it has to take responsibility and act on the global climate by ending its national oil and gas production.

Norwegian goodness as cultural commitment

Following Tsing, understanding the situated, sociocultural nature of scale-making requires connecting scales to central cultural commitments (Tsing 2000). In the following, I propose narratives of Norwegian goodness as a cultural commitment that both oil industry and climate activists orient themselves towards, albeit in different ways, when trying to assess what it is right and good for an oil-producing nation to do amidst an escalating global climate crisis.

In unfolding doing good as a cultural value, I approach Norway's oil experience not as a magnificent stroke of luck or the result of a particular skilfulness in managing the oil but as a narrative informed by a tradition of Norwegian goodness (Tvedt 2006; Witoszek 2011; Anker 2020) and Nordic exceptionalism (Loftsdóttir and Jensen 2012; see also Habel 2015; Diallo 2019; Khalid 2021). Scholarly discussions about Norwegian national identity and self-perception critically engage with the dominant cultural idea that Norway can somehow speak on behalf of what is understood to be "universally good" and holds a cultural commitment to be and do good (Tvedt 2006, 67; see also Anker 2020). The historian Terje Tvedt has labelled these cultural ideas a "national regime of goodness," highlighting the ambivalent nature of goodness as an imperative (Tvedt 2006). Specifically, Tvedt points out that the national regime of goodness hinders criticism and self-reflection by working to confirm the goodness of the system itself (2006, 69). The historian Peder Anker connects Norwegian goodness to environmentalism and Norwegian nature nationalism in a historical examination of Norwegians, who he calls "the world's green do-gooders" (Anker 2020, 238; see also Witoszek 2011). Anker argues that the foundation of this self-perception can be found among Norwegian environmentalists of the 1970s who glorified what they saw as pure Norwegian nature. Placing this pure and pristine nature as morally superior to dirty city centres by extension instantiated a hierarchy between a "beautiful, peaceful Norway contrasted with the polluted, troubled world" (Anker 2020, 4).

Discussions of Norwegian goodness are embedded within a larger framework of Nordic and welfare state exceptionalism. This literature challenges Nordic self-images that portray the Nordic countries as morally "good" because of their perceived non-involvement in European colonial activities and their universalist welfare systems, and it further explores the justice implications of such self-images, particularly for the dynamics of racialization in the Nordic region (Loftsdóttir and Jensen 2012; see also Habel 2015; Diallo 2019; Khalid 2021). These narratives of goodness are highly formative for how environmental and climate justice concerns are framed and deflected in a Nordic context.

Goodness narratives also pertain to national indigenous issues in Norway. The indigenous Sámi people, whose nation of Sápmi spreads across Northern Norway, Sweden, Finland and Russia, are deeply affected by both fossil resource extraction and renewable energy projects (see, e.g. Dale 2018), but Sámi issues often seem far removed from everyday life in Stavanger, which is located in South-West Norway. However, current contestations over wind turbines on Sámi land in Fosen, in Mid-West Norway, have started to generate more public awareness of landgrabs from the Sámi people in the name of the green transition. The turbines, erected with permission from the Norwegian state, were declared illegal by the Norwegian Supreme Court in 2021. The wind turbines, the court ruled, violate the rights of the Sámi population to practise their culture, as the turbines disturb Sámi reindeer herding practices in the area. The Norwegian state has refused to take down the wind turbines, causing massive protests from the Sámi community backed by Norwegian environmentalists and climate activists, all insisting the state should take down the turbines immediately (Børstad et al. 2021; Ballovara et al. 2023). The protests tell a story about a moment when the state's extractive practices cannot be reconciled with being good or moral despite the positive changes they are understood to have brought to Stavanger and Norway at large. This is a moment when cracks are forming in the idea of the state as a champion of goodness. And when cracks and frictions emerge in the dominant goodness narratives premised on the nation-state, this destabilizes the nation-state as a relevant and legitimate frame for assessing what is good and for whom.

When invoking the nation-state through the common good of the national community, or arguing that Norwegian oil can be good for the climate, the people working in the industry are not necessarily denying the existence or importance of dealing with global climate change. They are, however, using scales in a way that enables the status quo of continued Norwegian oil production despite its vast carbon emissions. This is a status quo that preserves the source of affluence and significance for their local community in Stavanger, the jobs the oil industry creates and the vibrant activity the industry brings to the city. I see the industry's foregrounding of the nation-state as a form of scale making that frames the relevance of oil in national terms, anchored in "the common good for the national community" as a central cultural value.

Interestingly, the young climate activists do not renounce the national regime of goodness. The aforementioned notion of a historical responsibility anchored in profit connects to Norwegian goodness in new ways. The activists understand Norway as particularly well-suited to set a good example and show that it is possible to end oil production. This well-suitedness is anchored in Norway's wealth, gathered in and reinvested through the oil fund, as well as what the activists see as a strong tradition of Norwegian welfare state democracy. Given its potential, Norway's failure to take the lead on phasing out its fossil fuel production is highly disappointing for the activists. This disappointment continues to fuel the activists' narrative of Norway as a nation that is supposed to do good and therefore must take the lead. Localizing global responsibility through insisting on the unity and inseparability of oil and carbon emissions from its extraction, production and combustion challenges the stability of the common good of the national community as a justification for continued oil extractions while still appealing to the goodness narrative. For the activists, sustaining goodness as a central cultural value requires that Norway must claim responsibility for the climate impact of carbon emissions from Norwegian oil and gas historically as well as presently, both in terms of national production emissions and emissions from the combustion of Norwegian oil and gas abroad. The activists thus challenge the prioritization of Norway's common good over the global common good, even if they continue to support Norwegian goodness as a central cultural value.

Goodness and comparisons to "bad others"

Scale-making also takes place through comparisons that place experiences and phenomena in relational fields, thereby enlarging or narrowing their relevance (Gal 2016, 91). In Stavanger, the nation-state is made relevant through comparisons to other oil states that emphasize Norwegian democracy and the social distribution of oil wealth through the welfare state, and occasionally through references to "clean oil."

Such comparisons, I suggest, are deployed to understand Norway's position and responsibility as different from other oil states due to democracy, strong regulative practices and the ability to redistribute wealth from oil extraction through the welfare state. Such comparisons confirm Norway's success at harnessing oil wealth to the benefit of the national community. As scale-making devices, these comparisons cut up connections and relations in the world in a way that positions Norway as a success and other oil states as failures. The countries that people in Stavanger most commonly used for comparison were Nigeria, Russia and Saudi Arabia.

To draw out how comparisons between Norway and other oil states are part of a larger framework of racial capitalism, I read them up against Hannah Appel's (2019) work on modularity in offshore oil extraction in Equatorial Guinea. Appel looks at the extractive practices of US oil companies in Equatorial Guinea and explores what she calls "the licit life of capitalism" (Appel 2019, 6). Appel refers to the unjust practices and legal frameworks that distance capitalist practices from local contexts and sites of production to ultimately make products – like oil – appear "*as if* untouched" by the practices and histories of its production (Appel 2019, 21 original emphasis). Appel argues that offshore oil creates the industry fantasy that extraction and profit move without friction from extraction site to extraction site (Appel 2019, 20–21).

Appel's work draws attention to practices of disembedding and distancing that make highly unjust contemporary capitalist production possible. I find Appel's argument especially compelling when thinking through the narratives of Norwegian goodness, innocence and success. At first glance, localizing oil by emphasizing that its origin makes it somehow more moral seems to run counter to the argument that Appel builds about disembedding products from local sites of production. In Norway, the localization of oil matters, because it evokes both the well-established regulation practices of the Norwegian state and the local blood, sweat and tears that have gone into creating the current oil infrastructure and pride in the national know-how of how to tame the turbulent North Sea. However, upon a closer look, localizing Norwegian oil is a narrative that rests on a moral architecture akin to that which Appel outlines. In Appel's ethnography, economic theory about "the resource curse" works to justify the continued exploitation of states on the African continent and to bypass the local community in an enclave style of production (Appel 2019, 23–25). In Norway, claims about Norway's exception to the resource curse work to justify why Norway should continue to produce oil with reference to the capacity of the state to handle oil wealth in a way that benefits the local population and the national community.

Part of what Appel convincingly shows in her work on Equatorial Guinea and US oil companies is that there is a "mutually beneficial relationship between absolute rule and transnational oil" (Appel 2019, 8–9). More specifically, transnational oil companies could not work the way they do and produce cheap oil without the authoritarian regimes they support and keep in place (see also Mitchell 2013; Ghazvinian 2007). Places like Equatorial Guinea, or more commonly Nigeria or Saudi Arabia, are used in Norway as examples of "bad others," places where oil is produced in harmful ways that make Norwegian oil production seem good. I do not mean to suggest that the relatively safe working conditions, well-paid jobs and the positive national and local ripple effects that oil creates in Stavanger and Norway are not preferable to the violent extractive practices of transnational oil companies in Equatorial Guinea or Nigeria. What I do suggest is that Norwegian goodness hinges on the perceived badness of other oil states. And that sustaining this goodness currently rests on governance structures that enable a discursive decoupling of Norwegian oil from the climatic impact of its carbon emissions. Discursively decoupling oil and carbon emissions contributes to these comparisons between Norway and other oil states, where Norway can call itself better because the comparisons primarily concern how different oil-producing societies are structured and focus less on how their oil enterprises contribute carbon emissions to the common global atmosphere.

Concluding remarks: discounting climate responsibility through discursive decoupling

I began this chapter by asking what different scalar conceptions mean for how activists and the industry ascribe responsibility for the climate impact of Norwegian oil extraction and combustion. Answering this question led me through an analysis of the ideological underpinnings of scale and how scales can enact particular ontologies (Tsing 2000; Simons, Lis, and Lippert 2014). Concretely, I depicted how activists and the industry frame the relationship between the national and the global in different ways, which have consequences for how they perceive what is good and right for Norway to do as an oil-producing nation amidst an escalating global climate crisis. This led me to conclude that scaling through carbon enables both discounting responsibility and expanding it, depending on how the relationship between the national and the global is enacted. And particularly that the industry's discursive decoupling of source and carbon emissions enabled it to claim non-responsibility.

I showed that activists and the industry are both invested in the idea that Norway has a culturally anchored responsibility to do good. However, they disagree on what doing good entails. The industry places importance on how the national oil enterprise contributes to the common good of the national community through the redistributive and regulative mechanisms of the welfare state, making it seem preferable to extract oil in Norway rather than other oil states. This perspective rests on a separation of oil and the carbon emissions released from its extraction and combustion. I showed the productivity of analysing this phenomenon in terms of a discursive decoupling between oil and emissions. For the industry, oil is local and specific, whereas carbon emissions from oil are part of a global accounting that can be dealt with separately from oil itself. Scaling carbon as a global concern anchored in accounting schemes further enables the industry to discursively disconnect national oil production from the global climate, thereby avoiding localizing and accepting responsibility for the carbon emissions from Norwegian oil and gas. Discursive decoupling thus serves to discount climate responsibility and preserve the wealth, affluence and significance that oil affords locally and nationally.

Climate activists, conversely, refuse the industry ontology of (de)coupling and insist on the unity and inseparability of carbon emissions and their concrete, situated source. Through their insistence on unity and by placing importance on the source of carbon emissions, the activists claim that Norway holds responsibility for both emissions generated from oil production and for emissions generated from the combustion of exported oil. Hereby they also refuse the industry's discounting of responsibility and insist that Norway and Stavanger are places from where the global climate crisis can be acted upon locally by phasing out oil and gas production.

Acknowledgements

I am grateful to everyone in Stavanger who contributed to making this research possible by sharing their time and thoughts with me. Thank you to The Greenhouse Centre for Environmental Humanities at the University of Stavanger for hosting me as a visiting scholar during my fieldwork in Stavanger. I would also like to thank my co-editors and research partners in the SOCCAR project: Katinka Amalie Schyberg, Ingmar Lippert and Steffen Dalsgaard for their comments and insights.

Notes

1 This is a highly contested claim (see, e.g. Sunde 2016).
2 My dissertation focuses on generational differences between the young climate activists and their oil saturated local community (Lautrup 2022). This chapter partly builds on Chapter 1 of the dissertation, which analyses social stories about oil and goodness in Stavanger. The present chapter is complementary in that it analyses scale-making in relation to industry and activist understandings of climate responsibility, and how these understandings are tied up with ontologies of (de)coupling or unity between carbon emissions and their source.
3 For examples of Norwegian offsetting schemes abroad, see for example, Hook and Laing 2022; Angelsen 2017 and Lounela this volume.

References

Angelsen, Arild. 2017. "REDD+ as Result-based Aid: General Lessons and Bilateral Agreements of Norway." *Review of Development Economics* 21, no. 2 (May): 237–64. https://doi.org/10.1111/rode.12271

Anker, Peder. 2020. *The Power of the Periphery: How Norway Became an Environmental Pioneer for the World.* Cambridge University Press. https://doi.org/10.1017/9781108763851

Appel, Hannah. 2019. *The Licit Life of Capitalism: U.S. Oil in Equatorial Guinea.* Duke University Press.

Ballovara, Mette, Line Alette Bjørnback Varsi, Lena Maria Myskog, Silje Hausen Myrseth, and Solvår Flatås. 2023. "700 dager siden Fosen-dommen – nå aksjonerer de igjen." *NRK*, September 11, sec. sami. https://www.nrk.no/sapmi/700-dager-siden-fosen-dommen-_-na-aksjonerer-de-igjen-1.16548787

Behrends, Andrea, Stephen P. Reyna, and Gunther Schlee. 2011. "The Crazy Curse and Crude Domination - Towards an Anthropology of Oil." In *Crude Domination: An Anthropology of Oil.* Berghahn Books.

Berge, Gunnar. 2014. "Da Oljå Kom." In *Stavangeren. Medlemsblad for Byhistorisk Forening Stavanger*. Temanummer: Stavangers Oljehistorie.

Blühdorn, Ingolfur. 2007. "Sustaining the Unsustainable: Symbolic Politics and the Politics of Simulation." *Environmental Politics* 16, no. 2 (April): 251–75. https://doi.org/10.1080/09644010701211759

Børstad, Johannes, Jan Rune Måsø, Kristi Kringstad, and Elisabeth Strand Mølstad. 2021. "Samers rettigheter ble krenket da vindkraftanlegg ble bygget på Fosen." *NRK*, October 11, sec. dk. https://www.nrk.no/trondelag/vindkraftutbygging-pa-storheia-i-trondelag-_-norske-samer-mener-strider-mot-urfolks-rettigheter-1.15685096

Boyd, Emily, Maxwell Boykoff, and Peter Newell. 2011. "The 'New' Carbon Economy: What's New?" *Antipode* 43, no. 3 (June): 601–11. https://doi.org/10.1111/j.1467-8330.2011.00882.x.

Dale, Ragnhild Freng. 2018. "Making Resource Futures: Petroleum and Performance by the Norwegian Barents Sea." [Doctoral Thesis, University of Cambridge] University of Cambridge.

Dalsgaard, Steffen. 2013. "The Commensurability of Carbon Making Value and Money of Climate Change." *Journal of Ethnographic Theory* 3, no. 1 (Spring): 80–98. https://doi.org/10.14318/hau3.1.006

Dalsgaard, Steffen. 2022. "Can IT Resolve the Climate Crisis? Sketching the Role of an Anthropology of Digital Technology." *Sustainability (Basel, Switzerland)* 14, no. 10 (May): 6109–. https://doi.org/10.3390/su14106109

Diallo, Oda-Kange Midtvåge. 2019. "At the Margins of Institutional Whiteness: Black Women in Danish Academia." In *To Exist Is To Resist: Black Feminism in Europe*, edited by Akwugo Emejulu and Francesca Sobande, 224–31. Pluto Press.

Equinor. n.d. "Elektrifisering av plattformer." Accessed August 28, 2022. https://www.equinor.com/no/energi/elektrifisering-av-plattformer

Foster, John Bellamy. 1992. "The Absolute General Law of Environmental Degradation under Capitalism." *Capitalism Nature Socialism* 3, no. 3 (February): 77–81. https://doi.org/10.1080/10455759209358504

Gad, Christopher, Casper Bruun Jensen, and Brit Ross Winthereik. 2013. "Practical ontology. World(s) in STS and anthropology." *Tidsskriftet antropologi* 67: 81–100.

Gal, Susan. 2016. "Scale-Making: Comparison and Perspective as Ideological Projects." In *Scale Discourse and Dimensions of Social Life*, edited by Michael Lempert and Carr E Summerson. Open Access E-Books. University of California Press. https://doi.org/10.1525/9780520965430

Ghazvinian, John Hossein. 2007. *Untapped: The Scramble for Africa's Oil*. Harcourt.

Gjerde, Kristin Øye. 2002. *"Stavanger er stedet": oljeby 1972–2002*. Norsk Oljemuseum.

Habel, Ylva. 2015. "Om at udfordre den svenske exceptionalisme - At undervise som sort." *K&K - Kultur og Klasse* 43, no. 119 (September): 75–102. https://doi.org/10.7146/kok.v43i119.22246

Hernes, Sigrid, Jonas Erraia, and Sveinung Fjose. 2021. "RINGVIRKNINGER AV OLJE- OG GASSNÆRINGENS AKTIVITET I 2019." *Menon Economics*.

Hook, Andrew, and Timothy Laing. 2022. "The Politics and Performativity of REDD+ Reference Levels: Examining the Guyana-Norway Agreement and Its Implications for 'Offsetting' towards 'Net Zero.'" *Environmental Science & Policy* 132 (June): 171–80. https://doi.org/10.1016/j.envsci.2022.02.021

Khalid, Farhiya. 2021. "Nordisk exceptionalisme & den hvide uskyld. Samtale med Elizabeth Löwe Hunter." Public Square. https://publicsquare.dk/artikel/nordisk-exceptionalisme-den-hvide-uskyld

Knox, Hannah. 2015. "Carbon, Convertibility and the Technopolitics of Oil." In *Subterranean Estates*, edited by H. Appel, A. Mason, and M. Watts, 309–24. Cornell University Press.

Krogh, Finn E., and Björn Lindberg. 2022. "14. juni 1972: 50 år med verdens beste oljemodell?" *Stavanger Aftenblad*, June 14. https://www.aftenbladet.no/meninger/debatt/i/47pLgo/14-juni-1972-50-aar-med-verdens-beste-oljemodell

Lautrup, Andy. 2022. "Generation Carbon - Goodness, Loss and Youth Climate Activism in Norway's Oil Capital." [Doctoral Thesis, IT University of Copenhagen] IT university of Copenhagen. https://pure.itu.dk/en/publications/generation-carbon-loss-goodness-and-youth-climate-activism-in-nor

Law, John, and Annemarie Mol. 2002. *Complexities: Social Studies of Knowledge Practices. 2. printing.* Duke University Press.

Lewis, Hilde Øvrebekk. 2012. "Julegaven ble til 13 millioner barnehageplasser." *Aftenposten*, May 27. https://www.aftenposten.no/okonomi/i/9vWoM/julegaven-ble-til-13-millioner-barnehageplasser

Lippert, Ingmar. 2011. "Extended Carbon Cognition as a Machine." *Computational Culture* 1: 1–17.

Loftsdóttir, Kristín, and Lars Jensen. 2012. *Whiteness and Postcolonialism in the Nordic Region: Exceptionalism, Migrant Others and National Identities.* Ashgate.

McNeish, John-Andrew, and Owen Logan. 2012. *Flammable Societies: Studies on the Socio-Economics of Oil and Gas.* Pluto Press. https://doi.org/10.2307/j.ctt183pbx9

Mitchell, Timothy. 2013. *Carbon Democracy: Political Power in the Age of Oil.* Verso.

Mol, Annemarie. 2002. *The Body Multiple, Ontology in Medical Practice.* Duke University Press.

Mol, Arthur P.J., and David A. Sonnenfeld. 2000. "Ecological Modernisation around the World: An Introduction." *Environmental Politics* 9, no. 1 (November): 1–14. https://doi.org/10.1080/09644010008414510

Newell, Peter. 2012. "The Political Economy of Carbon Markets: The CDM and Other Stories." *Climate Policy* 12, no. 1 (December): 135–39. https://doi.org/10.1080/14693062.2012.640785

Norgaard, Kari Marie. 2011. *Living in Denial: Climate Change, Emotions, and Everyday Life.* MIT Press.

Norges Bank Investment Management. 2022. "2025 Climate Action Plan." *Norges Bank Investment Management.* September 20. https://www.nbim.no/en/responsible-investment/2025-climate-action-plan/

Oil Change International. 2017. "The Sky's Limit Norway: Why Norway Should Lead the Way in a Managed Decline of Oil And Gas Extraction." https://priceofoil.org/content/uploads/2017/08/The-Skys-Limit-Norway-1.pdf

Ryggvik, Helge. 2010. *The Norwegian Oil Experience: A Toolbox for Managing Ressources?* Centre for Technology, Innovation and Culture (TIK-centre).

Ryggvik, Helge, and Berit Kristoffersen. 2015. "Heating Up and Cooling Down the Petrostate: The Norwegian Experience." In *Ending the Fossil Fuel Era*, edited by Jack P. Manno, Thomas Princen, and Pamela L. Martin. MIT Press.

Simons, Arno, Aleksandra Lis, and Ingmar Lippert. 2014. "The Political Duality of Scale-Making in Environmental Markets." *Environmental Politics* 23, no. 4 (March): 632–49. https://doi.org/10.1080/09644016.2014.893120

Sølvberg, Ingrid. 2022. "Oljehistorien bygger framtiden." *Stavanger Aftenblad*, December 6. https://www.aftenbladet.no/meninger/debatt/i/k6n61k/oljehistorien-bygger-framtiden

Strathern, Marilyn. 2020. *Relations: An Anthropological Account.* Duke University Press. https://doi.org/10.1515/9781478009344

Sunde, Inger. 2016. "Ren, norsk olje?" NRK. October 3, 2016. https://www.nrk.no/dokumentar/xl/ren_-norsk-olje_-1.13150883

The Norwegian Petroleum Directorate and The Norwegian Ministry of Petroleum and Energy. n.d. "Exports of Norwegian Oil and Gas." *Norwegianpetroleum.no.* Last accessed March 8, 2022. https://www.norskpetroleum.no/en/production-and-exports/exports-of-oil-and-gas/

Tsing, Anna Lowenhaupt. 2000. "The Global Situation." *Cultural Anthropology* 15, no. 3 (January): 327–60. https://doi.org/10.1525/can.2000.15.3.327

Tvedt, Terje. 2006. "Utviklingshjelp, Utenrikspolitikk Og Den Norske Modellen." *Historisk Tidsskrift* 85, no. 1 (March): 59–85. https://doi.org/10.18261/ISSN1504-2944-2006-01-04

Witoszek, Nina. 2011. *The Origins of the "Regime of Goodness": Remapping the Cultural History of Norway*. Universitetsforlaget.

Index

Pages in *italics* refer to figures, pages in **bold** refer to tables, and pages followed by n refer to notes.

accountability 5–6, 99, 101, 104–5
activism 188, 191; actor-network-theory 159–60, 178; climate activism 188; climate activists 185–86, 189, 194–97, 200; youth activists 195
agency 4, 18, 20n1, 177; agencies 175, 179; of carbon 2, 14; and structure 57
agricultural: fertilization 18; fields 70; frontier 143; interest organization 70; land 69–70, 123; practices 70, 74; production 6, 13; research institution 72, 77; sector 67–69, 71, 73, 75, 78; soils 65, 73; *see also* farming
agriculture 19, 60n3, 65, 67, 122–23; industrial 119
Anthropocene 14
Appel, Hannah 198
archive 17, 19, 158, 162–66, 171–78, 181n14

Barad, Karen 157, 160–61, 174, 180n7, 180n10
behaviour 8, 53, 57, 59, 117–18, 124, 128; change 10–11, 15, 18–19, 25, 46–48, 52–57, 59, 60n5; energy 28; environmental 47–48, 57; environmentally friendly 52; individual 11, 15, 47–48, 53, 56–57, 59; nudge theory of 55–56; rational choice theory of 52–53, 56; theories of 15, 52–53, 55–56
biochar 13, 15, 18, 65–78, 78n2, 79n6–79n7
Bourdieu, Pierre 7, 14
Bridge, Gavin 1, 5
Butler, Judith 104

calculation 2, 14, 46, 49–52, 54, 87, 97, 102, 105, 113–14, 125–27, 138; calculative assumption 50; calculative logic 129; carbon 113; carbon footprint calculators 15, 46, 52, 56; measurement 129, 138, 149; *see also* numbers
capitalism 11, 14, 113, 124; capitalist fixing 135; capitalist practices 198; capitalist system 13; climate 5; environmental effects of 114; racial 198; sustainable 193
carbon: accounting 6, 9–10, 13, 66, 68–70, 76–78, 87, 97, 127, 138, 193; budget 65; carbon dioxide removal (CDR) 65–66, 78n3; as commodity 12, 117, 133–34, 138–40, 147–49; compensation 66, 127, 148; credits 12–13, 16, 60n2, 75–76, 113, 136; dioxide 6, 65, 110, 125, 136; in everyday life 2, 5, 7–8; farmers 13, 68, 76–78; farming 15, 67, 71; frontier 16, 133, 135, 138, 140, 143, 147–50; governance 114, 117, 127; imaginary 113, 115, 134, 160, 167; logic 87, 112, 114, 119, 129; market 5, 12, 59n1, 76–78, 117, 134, 136–38, 147, 149; as material object 3; as metric 1–6, 57, 110, 113, 124; neutrality 11, 46, 71, 115; offsets 3, 47, 59, 60n2, 66–67, 70, 76, 111–12, 127–28, 139, 193; the social life of 133–35, 138, 140, 143–44, 147, 149; subjects 66–68, 71–72, 76–78; tax 47, 67, 70–73, 76; trading schemes 136, 192–93

carbon emission 2, 7; calculation 49–50, 59; from farming 68, 72, 76; individual 46, 54; mapping 97; national 86, 88, 96, 110, 192, 194; numbers 98–99, 103–4; from oil 187, 189, 191–95, 197, 199; as potential 9; reduction 5, 15, 25–29, 31, 40–41, 46–47, 66, 77, 88, 105, 113, 135, 149, 193; reduction projects 31; reduction schemes 27, 31, 41; sociocultural context of 29; sociocultural dimension of 26–27; source of 147, 200

carbon footprint 1, 7, 10, 18, 46–54, 56–59, 88, 95–106; calculator 15, 46, 48, 52, 56; displaying 16, 87, 100

carbon removal 66–73, 76–78, 78n2, 79n5; providers 68, 70–72, 76–77; sink 12, 60n2, 125, 133, 147

carbon storage 134–35, 138–39, 147, 193; carbon sink 4, 12, 60n2, 67, 110, 125, 133, 147, 193; store 13, 67, 71, 132–33, 135, 137, 193

Central Kalimantan 12, 16, 132

change: collective 56, 59; and continuity 2, 10–11, 89; cultural 2, 10, 14, 18, 26, 133; personal 10, 14, 25, 47, 53, 55–57, 59; social 2, 6, 10, 73, 78, 133–34; technological 10; *see also* transition

China 16, 110

Christianity 89, 101–2, 105

Church, the Danish People's 17, 86

citizen: citizen-led transformation 28; demands 3; engagement 29, 41; ideals 124; -state relations 114, 118

Clean Development Mechanism 12, 192–93

climate: activism 188, 191; credits 75, 77; justice 14, 17, 195–96; neutral 11, 71, 75; policy 65–67, 70, 73, 78; targets 66, 71, 187

climate change mitigation 3, 5, 11, 16, 25–26, 28, 46, 66–70, 72, 77–78, 79n6, 88, 97, 99, 101, 104, 132, 139, 143–44, 149; discourse 134; schemes 134

CO_2 *see* carbon

coal 15–16, 155–56, 172–73

collective action 47–48, 57, 59

collective self-consumption 34–37, 39

colonial 77, 141, 160, 196; neo-colonial climate relations 77

commodity: carbon as 12, 117, 133–34, 138, 140, 147–49; fictitious 139, 149; frictitious 134–35, 139, 147, 149; frontiers 138; peatland as 136

common good 17, 189–90, 192, 196–97, 199

community 15, 19, 120, 156; community driven environmental work 127; community-scale energy projects 26–30, 32–34, **35**, 37, 42; counter-cultural 178; eco-communities 15; energy 18, 25, 27–28, 30, 32, 42; engagement 32–34; green 48, 57–59; local 16, 126, 137, 198; moral 114; national 17, 189–90, 194, 196–99; participation 5, 28, 32–38; of workers 113

configuration 13, 15, 68, 73–74, 77–78, 148, 158, 171; user 68, 73, 78

conflict 16, 76–77, 103, 113, 126–27, 155, 177, 179

conservation 133–35, 176; agriculture 71; environmental 137; nature 176; nature protection 170; projects 133, 135

consumption 10–11, 15, 46, 48, 50–59, 97, 106; collective self-consumption 34–39; energy 7, 18, 25, 28, 43n1, 48–49, 155; patterns 51–53

continuity and change 2, 10–11, 89, 121

controversy 16, 86, 89, 93–94, 103

culture 1–4, 8–14, 19, 20n4, 49, 91, 159–60, 166–67, 172–73, 175, 180n5, 180n9, 196; change 1, 3; concept of 3, 8; cultural change/continuity 1–2, 10, 14, 18, 20n1, 26, 133; cultural commitments 185, 188, 195; cultural heritage 97, 103, 157–58; cultural values 16, 112–13; as meaning 3, 8, 12; *see also* socio-cultural

Dayak 12, 16, 133–35, 140–43

Denmark 49, 54, 65, 86, 187

discourse 3, 161, 173; civilization 114, 116, 118, 128; discursive decoupling 186, 192–95, 199; state 111–12, 118, 129

double claiming 68, 77–78, 79n9

ecological: awareness 116; collapse 128; compensation 173; degradation 19; destruction 173; ecological civilization 16, 110–16, 118, 120, 126–28; ecological consciousness 110, 118, 128; progress 115; restoration 118, 174; restitution 112, 118; redemption 110; relations 74, 78n3; socio-ecological system 149

economy 5, 9, 11; carbon 5, 9–12, 16, 67, 71, 77, 113–15, 117, 138, 154; economic assumptions 52; economic growth 110–12, 116, 193; economic imperative 75; economic incentives 10, 55–57, 72–73, 76; economic rationalities 8; economics 18; economic theories 47, 56, 59, 198; energy 171; low-carbon 5; national 187; rubber 144; techno- 26; value 7
elites 26, 29, 185
emissions: global 14, 67; national 67, 86, 187, 193; peak 115; scopes 51, 60n5; *see also* carbon
energy: communities 18, 25, 27–28, 30, 32, 42; companies or corporations 51, 162; consumption patterns 155; justice 28; poverty 26, 31, 33, 36–37, 39, 42, 43n1; poverty alleviation 26–29, 33, 42; projects 26–27, 38, 196; renovation **35**, 37, 40–41; use 25, 51, 96, 186
engagement: citizen 29, 41; participant 38; strategy **35**, 39, 41
environmental: accounting 87; behaviour 47–48, 52, 57, 59; change 57, 150; conservation 137; crisis 116, 134; damage 143; deed 54; destruction 113, 171, 173, 175, 177; experiment 121; extraction 134; governance 69, 79n12, 110, 115, 117; imaginaries 111; legislation 115; management 111, 155, 180n1; markets 189; politics 155; redemption 118; redress 111, 129; regulation 191; restoration 118; stewardship 116; values 113–14; work 113–14, 128; workers 111
environmental impact: of agriculture 60n3, 67, 138; of behaviour 60n5; of capitalism 113–14; of technology 66
environmentalism 25, 116, 125, 196; environmental activists 155, 167, 169; environmentalists 166, 196
ethics 26, 67, 111; ethical deliberation 114; ethical force 114; ethical impulse 124; moral duty 116
Europe 15, 25, 46, 65, 70, 86, 154, 185, 196
European Green Deal, the 42, 46
European Union 5, 28, 32, 60n2, 67, 71, 170
exemplarity 112, 128; good examples 14, 117–18, 197
experiment 117; agricultural 15, 60n3, 70, 75; decarbonization 75, 117; energy 27, 33, 76; environmental 119, 121, 123, 136; low impact life 56, 59, 155; policy 117–18; scientific 180n7; sites 59n1, 70, 75, 117, 119, 136; technological 15, 76; technocratic 117; *see also* pilot projects
experts 12, 25, 43n7, 73; expert knowledge 138; restoration 132, 138–46
extractivism 155, 178

farmers 15, 67–78, 79n8, 120, 128; carbon 13, 65–78; forest 134
farming 15, 67–74, 86, 92, 122, 188; forestry farms 113, 123, 127; *see also* agriculture
feminist technoscience studies 157
forest: afforestation 16, 65, 111–12, 114, 118–25, 128–29; carbon forestry 134; as carbon sinks 4, 6, 135, 140, 60n2, 193; deforestation 20n3, 133, 143; degradation 136; farms 113, 127; logging 136, 138, 143–44; planting of trees 55, 112–13, 119, 125, 128, 137, 144–45; production 144; REDD+ 6, 136–37, 143; reforestation 17, 60n2, 65, 86, 92; as sacred 141; value of 3, 13, 140; wetland 135, 142
fossil fuel 7, 46, 155, 186, 193, 197; replacing 76
France 15, 27, 30, 39
frontier 133–35; agricultural 143; carbon 16, 133–35, 138, 140, 143, 147–50; commodity 138; conservation 134; imaginary 134; property 141; resource 133–34, 140, 147, 150
future, the 9, 17, 66, 105, 110, 113–14, 116–17, 121, 123–24, 128–29, 154, 174

generations 17, 120, 123, 178; future 121–24, 143, 190; inter-generational 17, 124, 128, 200n2
German Democratic Republic 155
ghost 161, 174–76, 180n10
Girvan, Anita 7, 66–68, 72
global 1–3; climate governance 5, 117, 192; global justice 185, 195; and the local 30; markets 149; and the national 128, 185–86, 188–89, 194, 199; politics 66–67; responsibility 194, 197
Global North 3, 6, 67, 77, 155
Global South 6, 67, 185, 192–93
governance 10, 12, 33, 88, 93, 117–18, 138, 149, 190, 199; carbon 114, 127;

climate 5, 69, 150, 192; energy 28; environmental 69, 110, 115, 117; un-governance 88, 93
Graeber, David 8–9, 14, 20n4, 114
Great Green Wall of China 110, 118–20, 128
greenhouse gasses 1–2, 4, 6–7, 11, 46, 79n4, 110, 113; *see also* carbon
greenwashing 174
growth: green 116, 193; economic 110, 112, 116, 193
Günel, Gökce 6, 11, 68, 78

heritage: cultural 97, 103, 157–58; infrastructure 156, 158; politics 159, 161, 179
hope 47, 53, 95–96, 129, 175, 191, 193; for a greener future 128; planetary 128
Hoskin, Keith 105
household: energy-poor 15, 19, 25, 32–33, 40; low-income 26, 28, 30, 32; vulnerable 29–30, 32–33, 39, 42
human: actors 3, 12, 171, 186, 189; agency 4; carbon as a metric of 1, 57; geography 29, 160; human and nature 111, 116, 129, 173–75; humanity 116, 122, 128; humankind 14; and more-than-human 139, 177; and non-human 133, 149, 157, 186, 189

identity 19, 171, 177, 195
ignorance 154
imagination 26, 90, 154, 156, 159, 171–73; sociological 156, 159
indigenous people 12, 140, 180n8, 181n14, 196
Indonesia 12, 132–50
information 47, 52–57, 59, 95–96, 106, 145–46, 155–56, 162, 166–69, 171; carbon as 2; and change 1, 15, 47, 53–54, 56, 59, 96; extraction 171; as infrastructures of inscription 156; and reality effects 157
infrastructure: green 173; heritage 156, 158, 161; knowledge 175–76; restoration 149
inscription 156, 161, 172, 175–76; devices of 156, 161; infrastructures of 156, 161; technologies of 176
Italy 15, 27, 30, 38

justice 14, 17, 28–29, 66, 127, 185, 195–96; environmental 14, 17, 185, 195–96; global 185, 195

Kalimantan 133–45
kinship 111, 113, 123, 127–28, 142; extended families 126, 143; family members 124–26, 142; family ties 186; genealogies 142; obligations 123, 128; ties 113; *see also* generations
knowledge: of carbon 2–3; creation 139, 149, 157, 174; distribution of 14, 31; expert 132, 138; forms 3, 149, 161; infrastructures 175–76; and numbers 100, 102; politics of 154; practices 160; registers of 160; *see also* information
Knox, Hannah 5–7, 10, 46, 114, 193
Kyoto Protocol, the 5–6, 12, 192–93

land rights 133, 142–43; national 133, 142–43; private 142
landscapes 17–18, 121–22, 172; access to 133; as carbon sinks 12; change 16, 19; degraded 118; human intervention on 119, 147; mining 155; post-carbon 154; ruined 155, 173; and social relationships 142–43
Leach, Melissa 8, 20n1
lignite 16–17, 154–79
local: and the global 30; locality 34, 188; localization 198
Lohmann, Larry 138
Lovell, Heather 6
low-carbon lifestyles 4, 110, 118, 124, 128
Lusatia 157, 162, 164, 166–68, 176

MacKenzie, Donald 6
management 5–6, 8, 20n3, 39, 58, 66, 68, 72, 76–77, 95, 97, 106, 144, 155, 169, 173, 180n1, 187
Manchester 10
map 103, 162–66, 169, 171; carbon footprint 53, 106
market: carbon 5, 12, 59n1, 76–78, 117, 134, 136–38, 147, 149; voluntary carbon market 76–77, 138
materiality 154, 179, 20n4
material semiotics 20n4, 156–58, 160–61, 189
McLaren, Duncan 66
memorial 169, 176
metrics: carbon 1–6, 18, 57, 110, 113, 124, 129; *see also* calculation; numbers
mining: of coal 155–56, 172–73; of lignite 16, 154–79; open cast mining 155, 164, 166–67, 177; post-mining landscape 16, 155–56, 169, 172

mitigation deterrence 66
Mol, Annemarie 157, 160–61, 175, 180n6, 186
multi-species 139, 149; relations 139, 149; more-than-human entanglements 139; more-than-human life 177

national: carbon market 117; climate change mitigation 133, 137; climate change politics 147; climate goals 75, 77; climate policy 70, 87; climate target 71; community 17, 189–90, 194, 196–99; emission accounts 79n9; and the global 185–86, 188–89, 199; ideals 128; identity 19, 195; ideology 16; legislation 30, 33; narratives 17, 116, 190, 194; reduction goals 86, 88–90; reduction targets 149, 187, 196
nature: conservation 176; commodification of 134, 137–38, 147, 149; destruction of 178; understanding of 8, 116, 141, 159, 173, 196
nature-culture 175; naturescultures 158–61, 167, 172, 175–77, 179; separation of 8–9, 159–60, 173, 175
net-zero 2, 11; goals 65–66, 77; scenarios 65; targets 70
Newell, Peter 5, 8, 20n1
non-human 133, 149; actors 186, 189; entities 157, 172
norms 7–10, 19, 53–56; Social norms theory 55–56
Norway 3, 17, 86, 106n1, 134, 136, 150n2, 185–200
Norwegian Petroleum Museum 190
numbers 6, 12, 88, 97–105; and subjectivity 99, 101; *see also* calculation; metrics

offsetting *see* carbon
oil 3, 17, 185–200; combustion 17, 186–87, 195, 200; extraction 17, 187, 197–99; industry 17, 186–97; palm 136, 143; *see also* Norwegian Petroleum Museum
ontology 157, 159–61, 180n5, 186, 194–95, 199–200; extractivist 176; ontic 159–61, 174–75, 180n5; ontological politics 160, 170, 175; ont*-politics 17, 154, 157, 159, 161, 176–79
organization 86–91, 96, 105, 106n3; organizational values 87–88; organizational ethos 86, 90
Oudshoorn, Nelly 68

Paris Agreement, the 19, 46, 70, 134, 138
peasants 67, 124, 145, 148; *see also* farmers
peatland 12–13, 16, 132–49
pilot projects 30–31, 33, 40–42, 59n1, 136; *see also* experiment
Pinch, Trevor 68
plantations 136–37
policy: -driven change 15; experiments 117; government 115; implementation 15, 26, 41; initiatives 10; makers 9, 26; mobility 25–27, 29–30, 33, 42; national climate 65–67, 70, 73, 78, 133; public 112–13; of quantification 110, 128; restoration 136–37; state 118, 127, 129; studies 14
politics: carbon 114, 154; climate 66, 147, 156; environmental 155; global climate 66; of green transformations 20n1; national 113, 147; ontological 160, 170, 175; ont*- 17, 154, 157, 159, 161, 176–79; technopolitics 193
pollution 110, 113–14, 116–18, 122, 125, 154
Porter, Theodore 99–102
protest 156, 158, 165, 168–69, 171–72, 182, 196

quantification 16, 18, 110, 113, 125, 127–29

religious: doctrines 116; freedom 87, 90–92, 94, 101; values 113
renaturation 158
renewable energy 25–26, 31, 42, 119; communities 25, 27–28; projects 27, 196
resource: access to 7, 133; carbon 17–18; carbon storage 135; cultural 1–2; curse 188, 198; frontiers 134, 140, 147, 150; extraction 115, 141, 171, 174, 196; material 13, 178; natural 113, 133, 138; nature as 116
responsibility: climate 14, 185, 188; discounting 186, 199; environmental 128; historical 17, 185, 194, 197; individualization of 57, 59; locating 17, 126, 186, 194, 197; national 185, 192, 194–95, 197, 199–200; signaling 96–97, 99–100, 105; *see also* accountability
restoration: ecological 118, 174; environmental 118; land 17; peatland 12, 132–34, 139, 144, 147, 149

Sahlins, Marshall 7–9, 11–13, 20n4
scale 17–18, 41–42, 117, 185–86, 188–89; scale-making 17, 188–89, 195, 197; scale up 32, 42, 73, 75, 128
science and technology studies (STS) 6, 14, 156–57, 159–60, 178, 186
Scoones, Ian 2, 5–6, 8, 20n1
socio-cultural: dimensions of climate change 11; dimensions of carbon 26–27, 33–34, 40, 42; dynamics 25–26; imaginary 113
sociology 14, 156; sociological imagination 156, 159, 178
socio-material: configurations 148; landscapes 141; relations 133, 139, 143, 149
socio-technical 19; imaginary 111, 113, 115
soil 65, 69, 71, 119, 128, 142
solar energy: community 26–28, 30, 32–33; projects 25–43; pro-poor 18, 26, 41
Spain 15, 27, 30, 34, 36
standards 3, 6, 12–13, 67, 76, 102
state: state-church relations 15, 86, 91–94, 100, 104; state-citizen relations 114; welfare 17, 188, 190–92, 196–97
storying 154–55, 161
subject: carbon 66–68, 71–72, 76–78, 111, 114; green 2, 10; human 157, 174; position 125; responsible 104
subjectivity: carbon 2, 16, 129; and numbers 99
subjectification 6, 104; moral 105

tax *see* carbon, tax
technical adjustments 11, 68, 78
technology 74; and carbon emission reduction 26, 68, 73; carbon storage 193; of distance 100, 102, 105; focus on 68, 73; practice-technology 68, 73; technological innovation 67
territorialization 18, 133, 138, 147–49
transition 156; brown 177; energy 31, 33, 40, 42, 158, 179; green 2, 5, 47, 87, 95–96, 101, 103–4, 177, 179, 196; just 40, 43; low-carbon 40; politics of 157; regional 158; sustainable 58

transformation 1, 156, 158; brown 157–58, 173, 178–79; citizen-led 28; cultural 2, 19; green 2, 5, 20n1, 155, 158, 174, 178; of landscapes 154, 173; regional 158; transformational role of carbon 1, 3, 9; troubled 16; *see also* transition
trees 3, 14, 122, 139, 142, 147, 168; nursery 121, 126; planting of 60n2, 112, 119, 121, 124–26, 128, 132, 137, 144; *see also* forest
Tsing, Anna 134, 188–89, 195

UN, the 115, 134–36, 139

valuation 10; of carbon 3, 8–9, 11, 15, 147
value 8; anthropological concept of 8–9, 20n4; carbon as a value form 1–3, 9–10, 12, 14–17, 113–14; changes 11; conflicts 113; cultural 7–8, 11, 88, 90–92, 112–13, 116, 173, 195, 197; economic 7; hierarchy of 14; moral 8, 141; national 16, 19, 90, 92, 147, 149, 195–97; organizational 88, 90; quantifiable vs. non quantifiable 111, 126, 129; regimes 16, 140, 147–49
Verran, Helen 157, 160–61, 174–75, 180n8, 181n15
village 16–17, 112–13, 122–23, 128, 132, 135, 140, 143–44, 154–79; eco- 48, 58, 60n3; villagers 13, 16, 121–23, 132–33, 138, 141–47, *146*, 155, 158, 164, 168–68, 171–73

water 7, 16, 119–20, 122, 132, 135, 139, 141, 143, 145, 149, 167; waterways 132, 142–43, 145–46
welfare 3, 190; national 186; state 17, 188, 191–92, 194, 196–97
wetland 13, 132, 135–36, 140–43, 146, 149
wildfire 133, 135–36, 147
wind turbines 103, 167–68, 171, 196
Woolgar, Steve 68
Wuwei Prefecture 112, 118–20, 122, 126, 128

youth activism 185–86, 194–95